Global Environmental Governance, Technology and Politics

To my beloved children, Elias, Astrid and Ivan.

Global Environmental Governance, Technology and Politics

The Anthropocene Gap

Victor Galaz

Associate Professor in Political Science, Stockholm Resilience Centre, Stockholm University, Sweden

Edward Elgar

Cheltenham, UK • Northampton, MA, USA

Published by
Edward Elgar Publishing Limited
The Lypiatts
15 Lansdown Road
Cheltenham
Glos GL50 2JA
UK

Edward Elgar Publishing, Inc.
William Pratt House
9 Dewey Court
Northampton
Massachusetts 01060
USA

A catalogue record for this book
is available from the British Library

Library of Congress Control Number: 2013958019

This book is available electronically in the ElgarOnline.com
Economics Subject Collection, E-ISBN 978 1 78195 555 0

ISBN 978 1 78195 554 3 (cased)

Typeset by Servis Filmsetting Ltd, Stockport, Cheshire
Printed and bound in Great Britain by T.J. International Ltd, Padstow

Contents

Preface   The 'Anthropocene Gap'

The idea to write a book about governance and technology in a possibly new geological era dominated by humans, was born in May 2012 in the corridors of a gloomy and dark building in the industrial outskirts of London. Renowned Earth system scientist Timothy M. Lenton had just finished presenting a paper at a scientific conference on the likely transgression of an Arctic 'tipping point' in the next decades. In short, this implies that the North Pole could be largely ice-free in summer, inducing drastic changes in one of planet Earth's critical climate regulating levers with profound global consequences.

One could really feel the odd mixture between curiosity and nervousness amongst the audience. As we all left the room, I was immediately handed a colorful leaflet from what I thought, was a conference participant. After a closer look, I understood that the leaflet in fact was a call from a non-governmental organization called *The Arctic Methane Emergency Group* (AMEG). The group suggested putting a break on ice melting in the Arctic region, by reflecting away solar radiation through a geoengineering technology called 'cloud brightening'. In short, by deploying large fleets of ships with the ability to eject salt particles into the atmosphere, thereby brightening clouds and initiating a cooling of the areas.[1]

As I walked to the next session, it also became increasingly evident to me that there was no obvious scientific or political arena able to handle these intermingled scientific and social debates about potentially catastrophic Earth system 'tipping points', and suggested remedial but highly controversial technologies. Of course these issues were, and still are, debated in different international arenas – including the meetings of the parties of the Convention on Biological Diversity. But there really was no potent legal setting and international arena with the capacities to seriously investigate, discuss and govern these highly contested issues.

Then it hit me. This call by AMEG wasn't a unique and peculiar event at the fringes of sustainability science. On the contrary, it very much captured what I perceived and still perceive as the most critical challenges facing environmental politics and society in this new era of rapid environmental change: Earth system complexity and 'tipping points', technological change and the fragmented nature of governance in the Anthropocene.

As I elaborate in this book, the combination of these three issues creates a whole new set of institutional challenges that we as social scientists (or more precisely, political scientists) have just started to come to grips with.

This last claim might seem overstretched considering the long history of groundbreaking studies of global environmental problems and their institutional dimensions, championed by prominent social science scholars such as Oran Young, Nobel Laureate Elinor Ostrom and Frank Biermann. As I elaborate in-depth in this book, the political and institutional implications posed by the Anthropocene run deeper than currently has been acknowledged in current debates. The difficulties we all – citizens, scholars, decision-makers, business leaders and a flurry of non-governmental actors – have in grappling, analysing and responding to these issues, is what I explore as the 'Anthropocene Gap'.

WHAT THIS BOOK IS NOT

Global environmental change, emerging technologies and politics are issues that could very well fill a whole library. My ambition is obviously more modest. This is not a book about climate or biodiversity politics, environmental policy, or governance for sustainable development in general. Nor is it an analysis of 'green' technologies such as solar power; of how social media can support environmental awareness; nor an attempt to settle the debate between so-called techno-utopians and promoters of environmental doomsday scenarios. Instead, this is essentially a book about new institutional and political challenges posed by the interplay between rapid nonlinear global environmental change and emerging technologies.

This might very well be considered an extremely narrow perspective for such contested and multifaceted issues, and I agree. Current discussions about sustainability, politics and technology are immensely rich thanks to the vigorous long-term commitment from fellow scholars, entrepreneurs, activists, politicians and others. Whenever the analysis here feels too limiting, I would modestly urge the reader to keep in mind that institutions matter. That is, humanly devised institutions, and the way we organize the interplay between state and non-state actors (what I call *governance*), have repeatedly been proven to play a fundamental role in shaping, and responding to environmental change.[2] This insight applies all the way from locally contrived rules to govern forests, to global commons such as climate change and the ozone layer. Hence despite its limited focus, this book should be viewed as a contribution to intense and ongoing debates about how humanity is to navigate environmental change of an unprecedented scale and complexity. My approach to governance analysis is both analytical and normative. By that I mean that I combine an empirical and theoretical understanding of societies' capacities to steer environmental change, with a normative ambition to bring out shortcomings, and possible ways ahead (Dingwerth and Pattberg 2006).

As some readers might notice, I take on this task inspired by what some have denoted a 'resilience lens', that is, a focus on the ability of institutions and governance to grapple with change, surprise and multiple interactions between human–environmental systems (Gunderson and Holling 2002, Folke et al. 2005).

COMPLEXITY AND CONNECTEDNESS

The need for the social sciences to critically explore the political and institutional implications of rapid environmental change is urgent. Well-known terms like 'limits to growth', 'the great acceleration', 'planetary boundaries', 'a planet under pressure', and 'a new geological era' have one important thing in common: the attempt to capture the vast challenges posed by interacting global environmental stresses and a new proposed (in other words, debated) geological epoch on a planet fundamentally shaped by humans – the Anthropocene.

Despite an increased interest in these challenges and this proposed new geological epoch, we know surprisingly little about its implications for current debates on institutions and global environmental change. And scholars of environmental governance are only at the very beginning of grasping these deep repercussions.

This book presents and elaborates one key hypothesis: we are in the midst of an 'Anthropocene Gap', that is, a time where we are unable to grapple, analyse and respond to the major implications induced by our transgression into a human-dominated planet.[3] These three interrelated gaps can be summarized as follows: our *mental models and causal beliefs* (Lynam and Brown 2011) are being seriously challenged by the complexity, scale and speed of global environmental change; our *analytical approaches* (and here I focus on political science) are increasingly failing us as we gain increasing insights about the anatomy of Earth system change; and as a result, our *political institutions* at multiple levels of social organization are unable to effectively respond to novel risks and opportunities induced by interacting environmental, political and technological change. It is a bold statement, I know, and I will return to these acclaimed gaps explicitly in the synthesis chapter.

In this book I combine theoretical work, with in-depth analysis of four case studies. The cases range from the governance of (1) 'planetary boundaries', (2) geoengineering, (3) emerging infectious diseases, and (4) algorithmic trade in financial and commodity markets. While these might seem like very different issue areas, they all illustrate complementary aspects of critical, yet poorly understood institutional and political challenges posed by complexity and connectedness in social–ecological or human–environmental systems (to be defined in the next chapter).

STRUCTURE OF THIS BOOK

This book consists of two main parts. The first part is mainly theoretical and looks at current debates on global environmental change and complexity from a governance perspective. Chapter 1 Planetary *terra incognita* is an introduction, but also a summary of the critique of the notions of the Anthropocene and 'planetary boundaries'. I will not only put these terms in the context of similar notions such as the 'great acceleration' and 'limits to growth', but also provide a summary of current scientific and policy debates of the concept's validity and practical usefulness. The chapter concludes with my position in this scientifically and politically contested area.

In Chapter 2 *Governance and complexity* I explore key properties of complex systems that are of relevance for governance scholars. Here I try to present an overview of the governance challenges posed by complexity with a special emphasis on thresholds or 'tipping points'. This chapter also includes a synthesis of multidisciplinary insights on how social actors – ranging from policy-makers to artificial agents – perceive and respond to threshold behavior in human–environmental systems. I also link advances in 'early warnings' of pending catastrophic shifts in ecosystems, with some theoretical implications for governance (for example, early warning and response challenges). The chapter ends with a presentation of three 'governance puzzles', which will conclude the synthesis chapter of the book.

The second part of the book consists of in-depth analyses of four different case studies. Chapter 3, *Earth system complexity* discusses recent attempts by Earth system scientists to define a 'safe operating space' for human activity at the Earth scale. These so called 'planetary boundaries' are nine, possibly nonlinear Earth system processes that in addition to climate impacts, include ozone depletion, atmospheric aerosol loading, ocean acidification, global freshwater use, chemical pollution, land system change, biodiversity, and global nutrient cycles. In this chapter, I elaborate key international governance challenges posed by the notion of 'planetary boundaries', some emerging political tensions, misunderstandings, and some constructive ways to analyse these from a governance perspective.

The chapter also includes an elaboration of how global organizational networks of various forms attempt to respond to global 'tipping point' behavior.

Chapter 4 *Epidemics and supernetworks* instead focuses on the complex institutional and governance challenges posed by emerging infectious diseases (EIDs) such as animal influenzas (for example, 'avian influenza' and 'swine flu'), and hemorrhagic fevers such as Lassa fever and Henipah virus. While this might sound like an odd focus for a book on environmental change and complexity, it should be noted that diseases such as these are driven not only by increasing connectivity through trade and travel, but also by environmental factors such as land use change, climate change and rapid urbanization. In this chapter I explore how international actors such as the World Health Organization try to grapple with epidemic surprise in terms of early warning and response. Here I also explore the role information and communication technologies play in the way international actors collaborate across cross-national networks, and how these networks interact with more formal institutions such as the International Health Regulations.

Suggestions of large-scale technological interventions to combat climate change that a decade ago would have been discarded as science fiction are slowly moving toward the center of international climate change discussions, science, and politics. Chapter 5 *Engineering the planet* elaborates the intriguing governance challenges created by the development of geoengineering technologies – another illuminating example of the 'Anthropocene Gap'. The emphasis here is on the intricate governance challenges posed by emerging and converging technologies as we enter a new geological epoch. I explore regulatory gaps and the complex actor constellations in this domain, as well as the poorly understood and contested trade-off between innovation and precaution in a new setting characterized by rapid and nonlinear environmental and technological change.

In the last case study chapter (Chapter 6), I analyse another emerging technology with implications for our ability to govern global change in the Anthropocene: algorithmic trade in commodity markets. While market-based conservation policies, and the 'neoliberalization' of natural resources has already induced considerable academic debate (Arsel and Büscher 2012), the approach in this book is different, and focuses more on the dynamics of financial–ecological connectedness and their underlying technologies. Algorithmic trading (sometimes denoted as 'automatized trade', 'high frequency trade', 'computer based trading' or 'robot trade') is having profound impacts in the way and speed in which financial assets are traded. The capacities of computer algorithms to process increasing amounts of market information including financial news items, and conduce extremely rapid and complex trading patterns are clearly on the rise. As I intend to show, the rapid advancement of algorithmic trade pose until now unexplored environmental governance challenges due to the increased connectivity between financial markets, commodity markets, and ecosystem services on the ground.

The last chapter sums up the whole book and tries to show how we can start bridging the 'Anthropocene Gap'. It draws together key common conclusions across the cases, and links back to emerging theories on governance for sustainability in the Anthropocene. Hence this chapter summarizes theoretical insights related to the ability of governance – including institutions and networks at multiple levels – to cope with human–environmental complexity and connectivity at multiple temporal and spatial scales.

A THEORETICAL CONTRIBUTION

The issues elaborated in this book hopefully draw broader interest than for scholars of environmental governance. Governments constantly struggle to reconcile the need for institutional stability and flexibility through collaboration, institutional innovation and soft-steering instruments. Charles Perrow's (1984) now classic book *Normal Accidents* provides a detailed elaboration of the generics of complex technological systems and the type of organizations able to cope with their associated risks (see also literature on 'High Reliability Organizations'). Moreover, governance scholars such as Jan Kooiman, Jon Pierre and Guy B. Peters, present interesting insights related to the ability of governance systems to cope with change and uncertainty. Researchers following the innovative path laid out by the late Elinor Ostrom, have also shown an increased interest in the role of polycentric governance for more flexible and robust forms of steering in complex settings.

The issues explored in this book – such as coordination in multi-level networks, institutional flexibility, diversity and robustness – hence are strongly linked to governance analysis in general (for example, Pierre and Peters 2005, Kooiman 2003). This book therefore aims to contribute to this wider (in a sense non-environmental) theoretical debate, identify strengths and weaknesses in our understanding, and build an argument firmly anchored in rich case studies.

The issues explored here are several. What characterizes international institutions able to detect and respond to 'global human–environmental surprises' of great importance to human well-being? Are international institutions able to address complex Earth system interactions? And is it at all possible to create rules that are strong enough to 'weed out' technologies that carry considerable ecological risk, but still allow for novelty, fail-safe experimentation, and continuous learning? These are far from easy questions, and my intention is not to present simple answers, robust hypothesis testing or quick-fix solutions. My ambition instead, is to portray an extremely exciting evolving landscape of emerging issues, puzzles, and controversies at the very heart of debates on global environmental change, politics and technology.

Acknowledgements

This book would have been impossible without the generous support from a number of organizations, funders and colleagues all over the world. I'm indebted to the Resilience Alliance, the Earth System Governance Project, the Beijer Institute of Ecological Economics, and Stockholm Resilience Centre in particular, for their continuous efforts to provide arenas where scholars from different scientific disciplines meet and explore critical global sustainability challenges. Support from the Swedish *Futura Foundation* has been critical for the realization of this book – thank you for allowing me to bring all those messy ideas and mini-projects into one (hopefully) coherent book.

A number of colleagues have also contributed with important input as co-authors over the years. I would especially like to thank Andreas Duit, Beatrice Crona, Frank Biermann, Albert Norström, Henrik Österblom, Per Olsson, Örjan Bodin, Johan Gars, and Björn Nyqvist. Many of the chapters have been improved thanks to the much-needed comments on early chapter drafts provided by Melissa Leach, Will Steffen, Michael Schoon, Ralph Bodle, Larry Madoff, Stephan Barthel, Lennart J. Lundqvist, Gunilla Reischl, Rak Kim, Anne-Sophie Crépin, Maja Schlüter, Juan Carlos Rocha, Emilie Lindkvist, Jerker Thorsell, Jorge Laguna-Celis, and Manjana Milkoreit.

A number of people from international organizations (such as the World Health Organization and the Food and Agriculture Organization), non-governmental actors (such as Médecins Sans Frontières), and governmental bodies (such as the Swedish Environmental Protection Agency) have contributed with their valuable time in interviews. I can't mention them all here, but their help should be recognized.

I would like to mention a number of colleagues who have provided invaluable inspiration, and pushed me towards steep learning curves over the years. Will Steffen opened my eyes to the Anthropocene challenge in an evening lecture many years ago, and drastically changed the trajectory of my research plans as a young political scientist. Elinor Ostrom, Carl Folke, Johan Rockström, Thomas Homer-Dixon, Per Olsson, Frank Biermann and Frances Westley have been critical as sources of intellectual inspiration over the years. Per Olsson has also, as a close friend and colleague, made my life as a scientist considerably more fun that it normally would be. Carl Folke – I wouldn't be the sort of multidisciplinary scientist and optimist I am today without his energetic support and leadership over the years at both CTM and the Resilience Centre. The 'Wolfpack' (you know who you are), and the brilliant DJs, artists and musicians at SoundCloud.com have constantly injected me with positive energy, and made the last writing year (almost) a pleasure.

Fredrik Moberg at Albaeco and the Stockholm Resilience Centre deserves a special mention here. Fredrik has continuously read, reviewed, and commented on all chapters in this book. His help and always positive support as a friend, colleague and co-author has been truly invaluable throughout the whole writing process. And to my beautiful wife Karin Wettre – thank you so much for putting up with my travels, late writing evenings, and occasional mood swings during the last critical month. This book would never have existed without your love and unconditional support.

1. Planetary *terra incognita*

The year 2016 will be critical for the history of planet Earth. This is the year when the International Geological Congress will meet to finally settle the debate of whether humanity formally has entered a new geological epoch: the Anthropocene. This might seem like a superfluous subject for a scientific meeting to discuss considering the explosion of the concept in current policy and scientific debates. Not only did *The Economist* and *National Geographic* already in May 2011 produce special issues on this new era; in 2012, the British Broadcasting Corporation (BBC) broadcasted a series of documentaries on the 'human epoch', and publishing giant Elsevier inaugurated the new journal *Anthropocene* in 2013. Yet, the scientific debate has not been settled. And similarly contested concepts attempting to define humanity's impact on Earth – such as 'the great acceleration' and 'planetary boundaries' – are widely circulated amongst academics, concerned non-governmental organizations and policy-makers.

Any institutional and political analysis of global environmental change on a human-dominated planet, should build on a firm understanding of these concepts and their associated scientific and political debates. As I will elaborate, these disputes are becoming increasingly intense and difficult for outsiders to grapple. The reason I believe is simple: as research insights from the Earth system sciences gradually propagate through media and policy discussions, they renew existing environmental political controversies. This time, the debates are not only the familiar ones, such as the contested tension between economic growth and sustainability. Instead, they have a new focus on Earth system complexity, and unprecedented trade-offs in time and space.

In this chapter, I briefly summarize what some have called the 'Anthropocene debate', as well as current contentious discussions about the role of Earth system science and 'planetary boundaries' in political decision-making. I also discuss the critical role that perceptions about technological change play in this debate, and identify three overarching governance challenges (or 'puzzles') that I intend to explore in the concluding part of the book.

As I intend to show, the interesting question is not whether a new human-dominated geological era is formally here, nor whether the proposed transgression of 'planetary boundaries' should be reframed as 'planetary opportunities'. The truly exciting questions emerge as we try to unpack the novel institutional and political challenges that surface as humanity increases its domination over a complex Earth system.

THE ANTHROPOCENE DEBATE

Global environmental change has been on the international political agenda for decades. Some would trace it back to the first UN-led international meeting on sustainability in Stockholm 1972. As Robert Boardman argues, the origins of scientifically grounded studies of the Earth system can be traced to the eighteenth century, and especially the transformative developments in geology as a scientific discipline (Boardman 2010, p. 57).

Earth system sciences has always influenced, and been influenced by, broader social and ontological debates. In short, political, religious, cultural, institutional and other societal factors, substantially shape and frame perceptions about the Earth system (Boardman 2010, p. 71). There is no reason to believe that current notions of the Anthropocene and associated concepts such as 'planetary boundaries' are an exception. Uhrqvist and Lövbrand (2009) explore these issues in an interesting Foucauldian analysis, suggesting that Earth system science is not only a scientific endeavor, but should also be viewed as playing a key role in knowledge production and therefore in 'the formation of governmental practices' (p. 3). The Earth system science community, through its methodologies, international research programs and technologies 'has made the Earth System seem stable, comparable and diagnosable and hereby open for government intervention' (p. 21).

Processes of knowledge production hence matter. My position in this discussion is different, and my main argument is that these debates should not lead us to believe that we can overlook the institutional and political implications of the human enterprise entering the Anthropocene epoch. Will Steffen, Paul Crutzen and John McNeill (2007) sum the state of knowledge elegantly:

> The term [. . .] suggests that the Earth has now left its natural geological epoch, the present interglacial state called the Holocene. Human activities have become so pervasive and profound that they rival the great forces of Nature and are pushing the Earth into planetary *terra incognita*. The Earth is rapidly moving into a less biologically diverse, less forested, much warmer and probably wetter and stormier state. (p. 614)

The notion of the Anthropocene is often traced back to earlier talks and papers by Paul Crutzen (for example, Crutzen 2002), but the proposition that humankind's activities are so large to effectively shape the function of the Earth system, was raised by George P. Marsh as early as the 1860s. Steffen and colleagues (2007, p. 617) suggest that human imprints on the Earth system intensified drastically after the Second World War, triggered by rapid population growth, multiple technological breakthroughs and a growing world economy. The results of these changes can be seen clearly today: in the rate of species extinction, increasing ocean acidification, rapid deforestation, exponential releases of greenhouse gases like carbon dioxide and methane, and global modification of freshwater, mineral and nutrient flows. Just to mention a few.

These might all seem like well-known and discussed global environmental issues, especially since the intense climate change debates in the early 2000s following from Al Gore's blockbuster documentary *An Inconvenient Truth* (2006), Sir Nicholas Stern's renowned review on the economics of climate change (2007), and the Nobel Peace Prize acknowledgement to the Intergovernmental Panel on Climate Change (2007). But the implications of the Anthropocene argument are much deeper than a familiar sense of increasing environmental stresses on planet Earth. In my mind, it is a profound insight of how human activities are shaping the Earth system.[4]

TRANSFORMING THE PLANET

Four considerably less known examples are illuminating illustrations of how the human enterprise is shaping Earth in astonishingly profound ways.

In 2011, Miller and colleagues published a highly controversial estimate of the global availability of wind power as limited by the kinetic (= motion) energy inherent in the Earth system (Miller et al. 2011). The novelty in this paper is its attempt to link climate impacts to wind-power extraction. Essentially, the paper calculates the unintended climate effects of proposed large-scale (17 TW) deployment of wind technologies. The calculations contain an interesting observation stylishly captured by the *New Scientist*: 'He [Miller] concludes that it is a mistake to assume that energy sources like wind and waves are truly renewable. Build enough wind farms to replace fossil fuels, he says, and we could seriously deplete the energy available in the atmosphere, with consequences as dire as severe climate change'.[5]

Human impact on biological evolution is another example. The evolution of multi-drug resistant pathogens is one of the clearest examples of how human action – through the misuse and overuse of antibiotics – fundamentally changes evolutionary pressures on micro-organisms. Our impact on the rate and course of biological evolution applies across a range of species, and on global scales including biodiversity and ecosystems.

Powerful human-induced selection pressures on species unfold through, for example, climate change as species shift geographical ranges, and compete with new species (Norberg et al. 2013, Skelly et al. 2007). In addition, humans also induce 'non-natural' exploitive selection pressure by harvesting species and individuals of certain size, morphology and behavior, thereby leaving behind 'the less desirable to reproduce and contribute genes to future generations' (Allendorf and Hard 2009, p.9987). These human-induced selection pressures on species not only has genetic implications (for example, causing aquatic animals such as fish to 'shrink' due to warming and fishing pressure, Forster et al. 2012), but could also potentially influence ecological dynamics on a global scale (Palkovacs et al. 2012).

Earth system scientists had also made another uncomfortable discovery. Human emission of aerosol particles into the atmosphere – through for example the combustion of fossil fuels containing sulfur – are currently masking some of the warming we would otherwise see (Steffen et al. 2007, p.619). As many have noted, international policies to remove these health-threatening aerosols from the atmosphere, could potentially lead to rapid *additional* warming up to +1°C (Steffen et al. 2007, p.619, uncertainties are considerable, as some particles also contribute to warming, see Andreae et al. 2005 for a summary).[6]

Lastly, in 2009 a group of 28 scientists suggested that planet Earth's capacity to support human life and well-being is not only fundamentally dependent on climate regulation, but also on eight additional Earth system functions formulated as 'planetary boundaries' (elaborated in detail in Chapter 3). These 'boundaries' are nine, possibly nonlinear Earth system processes that manifest themselves at the planetary level and include in addition to climate impacts, ozone depletion, atmospheric aerosol loading, ocean acidification, global freshwater use, chemical pollution, land system change, biodiversity, and biogeochemistry (Rockström et al. 2009a, b). These 'boundaries' are not fixed, but instead rough estimates of how close to an uncertainty zone around a potential threshold that the global human community can act, without seriously challenging the continuation of the current state of the planet.[7] Bluntly put: transgressing these suggested 'boundaries' could risk pushing the Earth system into a whole new equilibrium with unknown consequences for human life on the planet (see also Barnosky et al. 2012).

This recent understanding on the scale and implications of the human enterprise on biodiversity, ecosystems and the Earth system as a whole might seem overwhelming. And as I discuss below, none of these results by the scientific community translate effortlessly into policy-making, current political debates and institutional options. Two aspects complicate matters considerably. The first are the ongoing scientific debates about the notion of the Anthropocene, and associated concepts such as 'planetary boundaries'. The second is an emerging political debate about the usefulness and implications of the concept.

THE ANTHROPOCENE AND PLANETARY BOUNDARIES – THE SCIENTIFIC DEBATE

Have we really entered a new geological era? A quick Google search and exploration in academic databases might give the impression that the concept is well-established, yet the academic debate is far from formally settled. The 'official gate-keeper of the geological time scale' (Kolbert 2010) is neither the buzz in international media, nor the volume of academic literature using the term. Rather, it is the International Commission on Stratigraphy (ICS) headed by geologist Jan Zalasiewicz at the University of Leicester (UK).

ICS regularly debates how to name different geological time periods (Kolbert 2010). For years, geologists debated whether the Quaternary – the most recent geological period commencing 2.5 million years ago – ought to exist at all. The critical issue in this geological debate is whether humans can be said to have contributed to 'stratigraphically significant change' (Zalasiewicz et al. 2008, see also Zalasiewicz et al. 2010), that is a clear global signature in rocks, soils and fossil records. Hence while humanity's imprint on the Earth system might seem overwhelmingly clear through climate change, biodiversity loss and land use changes, the strength of it has been questioned from a geological temporal perspective. Will scientists millions of years in the future be able to identify a distinct layer on the Planet's sediment that matches humanity's explosive imprint on the planet? And if that were the case, when exactly would this new epoch or era have started? In the year 1800 when human population hit its first billion, or 6000 years BC when the clearing of forests for agriculture and the expansion of irrigation for rice 3000 years later led to increases in atmospheric carbon dioxide and methane concentrations has been suggested to have prevented the onset of the next ice age (Ruddiman 2003, Steffen et al. 2007, p.615)? The (stratigraphic) jury is still out.

The results from their deliberations are likely to have broader societal impacts considering emerging discussions of how to understand the wider societal implications of a human-dominated planet. The reason is that the final decision on the birth of the Anthropocene affects how risky we perceive this shift into a new epoch to be.

Clive Hamilton, professor of public ethics, as an example, argues that proponents of an 'early Anthropocene' whereby humans have been a planetary force since the birth of human civilization, also implicitly suggest that there is essentially 'nothing fundamentally new about the last couple of centuries of industrialism. It is in the nature of civilized humans to transform the Earth, and what is in the nature of the species cannot be resisted. [. . .] the Anthropocene becomes in some sense natural' (Hamilton 2012, p. 3). Bluntly put: is this proposed geological transition an extremely risky anomaly in the history of planet Earth (cf. Steffen et al. 2011), or simply an extension of human societies' ingenious ability to engineer ecosystems to our species' benefit (Ellis 2012)?

As Jim Proctor (2013) notes, different and conflicting notions of what an Anthropocene future holds co-exist in current American environmental thought. At the center of this debate are different perceptions of 'nature' – what is the future of conservation and sustainability, if nature never was entirely natural, nor unnatural? And shouldn't we instead speak of the 'Obcene Epoch' to better reflect 'the layers of rubble that will pile up during the extinction of most of the plants and animals of the Holocene – the ruined remains of so many of the living beings we grew up with, buried in human waste' (Dean Moore 2013).

Similar but more intense debates can be identified for the notion of 'planetary boundaries'. For example, in parallel to the publication of the paper in *Nature* in 2009 (Rockström et al. 2009a), *Nature Climate Change* invited seven experts to comment on the definition and quantification of each individual boundary. While some responses where cautiously positive, others were skeptical to the suggested quantifications, to the possibility of threshold change, and to whether the boundaries were set at the appropriate scale (in other words, global). William H. Schlesinger for example, argued that the 'threshold for nitrogen seems arbitrary and might just as easily have been set at 10 per cent or 50 per cent', noting also that defining thresholds might lead to perverse incentives for policy-makers (Schlesinger 2009). Steve Bass (2009) raised similar concerns for the boundary for land use change, as did Myles Allen for the climate boundary (Allen 2009). Other criticisms related to the difficulties in defining clear global 'thresholds' for changes that have clear multi-level (global to local) and multi-system connections such as ozone depletion (Molina 2009), biodiversity (Samper 2009), and freshwater use (Molden 2009). The debate regained intensity at the end of 2012 after a paper by Brook and colleagues (2013). Their argument was – briefly put – that planetary-scale tipping points in the terrestrial biosphere were unlikely. The underlying reason is what they identify as a lack of strong interconnectivity – that is, changes in terrestrial ecosystems tend to remain localized or sub-continental events. Hence abrupt negative ecological changes, the argument goes, are not likely to cascade and add up to planetary-scale 'tipping points'. Terry Hughes and colleagues (2013) instead focus on the temporal dimensions of threshold changes, and argues that shifts between states can be so slow (decades, centuries or longer), that they create serious early warning and response challenges. In their own words:

> [W]e argue below that for systems that respond much more gradually than small lakes after transgressing a threshold, the challenge of identifying and avoiding tipping points is even greater, that is, ascertaining whether we have already crossed a threshold, are now living on borrowed time, and are shifting inexorably to a new regime. (p. 150)

Despite remaining scientific uncertainties, the attempt to define a 'safe operating for humanity' has led to an intense policy debate (elaborated below). Refinements of the identified boundaries have been suggested for phosphorus (Carpenter and Bennett 2011), freshwater use (Rockström and Karlberg 2010) and chemical pollution (Persson et al. 2013). Others, such as Steven W. Running (2012) have tried to identify additional measurable boundaries based on, for example, the terrestrial net primary production (NPP). Anthony D. Barnosky and colleagues' article 'Approaching a state shift in Earth's biosphere' takes a similar approach, and reviews the evidence for a potential abrupt change of the global ecosystem as a whole (Barnosky et al. 2012).

It should be noted that this is only a brief summary of scientific discussions within the natural sciences. Several attempts have also been made by social scientists to explore the potential implications of 'planetary boundaries' for institutions and governance, including global environmental governance (Galaz et al. 2011b, Galaz et al. 2012a, b, Biermann 2012, Schmidt 2013), and national policies (Nilsson and Persson 2012). These include more interdisciplinary approaches to explore the implications of 'planetary boundaries' for economic growth (Steffen et al. 2011b), and suggestions to reframe Earth system challenges as 'planetary opportunities' to bring out the adaptive and innovative capacities of societies through technology (DeFries et al. 2012, see also Westley et al. 2011).

PLANETARY BOUNDARIES – THE POLITICAL DEBATE

Political disputes about the precise temporal definition of the Anthropocene have been modest. The notion about Earth system scale 'planetary boundaries' – which are intimately linked to notions of a new human-dominated geological era – are a whole different story. While the original *Nature* paper (Rockström et al. 2009a) intentionally excluded a discussion of social drivers such as global distribution, trade, demographic change and technology in order to simplify the analysis, 'planetary boundaries' has been rapidly pulled into intense sustainability policy debates, and ideological controversies. Several examples highlight the heat in this recent debate.

Just as the world's nation states were to meet at the United Nations Conference on Sustainable Development (known as Rio+20) in June 2012, the US-based think tank *Breakthrough Institute* released a highly critical, and widely spread review of the 'planetary boundaries' framework (Blomqvist et al. 2012). The report questioned the underlying scientific evidence, the main results, as well as the claim that the transgression of the suggested boundaries would have detrimental implications for human well-being (Blomqvist et al. 2012, pp.4–5, see also debate in *Nature* between Lewis 2012 and myself in Galaz et al. 2012c).

Think tank-produced reports of this sort are overly common, but a successful spin in international media is not. The *Scientific American* (Biello 2012), *The Economist* (2012), and *The Wall Street Journal* (2012) all described the contents of the report. The timing was excellent from a lobbying point of view.

One of the topics on the Rio+20 agenda was a redrafted 'zero-draft' declaration, including an explicit reference to the need to stay within scientifically defined 'planetary boundaries'. This reference to the concept was later removed from the document, according to some of the negotiators involved, due to strong skepticism from amongst others the US, Chinese and G77 delegations. The main reason seems to be the perception of 'planetary boundaries' as quantified limits to growth. That is, the definition of a 'safe operating space' has, despite the authors' intent to focus on the prospects of human well-being on a planet experiencing rapid environmental change, reignited older political tensions about the need for environmental protection, and the right to development.

The perception of boundaries as limits has led to reiterated claims that 'planetary boundaries' are doomed to fail as an international policy framework as they are likely to create the same sort of political gridlock as experienced in international climate negotiations (Victor 2010). Such an argument overlooks several important developments.

One is the ambitious GEO-5 report coordinated by UNEP, and published in 2012 (UNEP 2012). GEO-5 is not only a scientific assessment of the state of the planet, but is also linked to parallel multilateral negotiation processes between member states whereby science is translated into policy-relevant messages. This translation is far from painless and requires considerable negotiations around definitions and formulations of key messages.

The summary for policy-makers in this case, included several statements that in fact summarize key messages from the 'planetary boundaries' synthesis, without using the actual term. The first section for example, is entitled 'Critical thresholds' and states: 'The currently observed changes to the Earth System are unprecedented in human history. [. . .] As human pressures on the Earth System accelerate, several critical global, regional and local thresholds are close or have been exceeded. Once these have been passed, abrupt and possibly irreversible changes to the life-support functions of the planet are likely to occur, with significant adverse implications for human well-being' (UNEP 2012, p. 6). At Rio + 20, nation states agreed on 'the continuation of a regular review of the state of the Earth's changing environment and its impact on human well-being and in this regard, we welcome such initiatives as the Global Environmental Outlook process aimed at bringing together environmental information and assessments and building national and regional capacity to support informed decision making' ('The Future We Want', #90). Hence while 'planetary boundaries' as a concept didn't make it into the final negotiated text at Rio +20, its main ideas in fact did: through a negotiation technical back door.

An additional debate (at the moment only in the blogosphere) has also emerged about the role of scientific advice on planetary boundaries, and associated proposals for global institutional reform. Roger Pielke Jr. (Professor at the University of Colorado at Boulder, as well as senior fellow at the *Breakthrough Institute*) argues that the notion of 'planetary boundaries' is associated with one political philosophy: the need for top-down interventions where 'issues of legitimacy and accountability are easily dealt with through the incontestable authority of science'.[8] Melissa Leach (Professor at the Institute for Development Studies, UK) as a precursor to this debate, also noted that simplified notions of planetary boundaries and interpretations emphasizing global scale urgency and crisis, have a tendency 'to align rather neatly with approaches that are top-down not bottom up, set rather than deliberated, singular rather than respectful of diversity, privileging scientific over experiential expertise, global rather than local, control rather than response-oriented, and so on'.[9] As I elaborate in this book, as well as in a longer blog response, I believe these to be highly simplistic notions of the governance implications of planetary boundaries.[10] But Leach raises an important point worth restating, and discussing further: advocates of particular forms of authoritarian governance can use selective interpretations of 'planetary boundaries' to advance their interests. Hence it becomes essential to acknowledge, and maintain a diversity of governance perspectives (Leach 2013).

Others have instead attempted to build on, rather than question, the framework. Kate Raworth from the UK-based non-governmental organization *Oxfam* expands the 'planetary boundaries' framework to include social dimensions such as human security, social equity, and gender equality (Raworth 2012). Mark Lynas's book *The God Species* is an additional contribution to the debate with a strong emphasis on the role of technologies (such as nuclear power, geoengineering and genetic engineering) as a means to stay within a 'safe operating space' (Lynas 2011).

This is only the tip of the iceberg in terms of different political interpretations of the notion of a new human-dominated geological era, and 'planetary boundaries'. Several international policy initiatives have already taken off, ranging from UN-led panels and EU-documents, to meetings between religious leaders under the auspice of the Dalai Lama (Galaz et al. 2012b). These initiatives are paralleled by critical reflections, such as this quote from a blogpost by Michael Shellenberger and Ted Nordhaus (both at the *Breakthrough Institute*):

> If the United Nations were seeking to further alienate the rapidly-growing developing world from common ecological action, it could do no better than to embrace this hoary, unscientific Malthusianism. [. . .]. The last thing the UN should be telling the four out of ten human beings who rely on wood and dung as their primary power sources is that we've reached the limits to human development, and technology can't save you.[11]

Put bluntly, even though many Earth system scientists would probably very much like to, it is practically impossible to decouple Anthropocene science and 'planetary boundaries' from its electric political context.

WHAT DO WE MAKE FROM THIS?

As the quote above illustrates, debates about the Anthropocene and proposals to keep human development within 'planetary boundaries' have, despite important differences in theoretical assumptions and methods (Rockström et al. 2009b, appendix 1, pp. 4–5), clear political similarities to older disputes between Malthusians and their antagonists; controversies about the 'Limits to Growth' in the 1970s; and discussions on governance options for sustainable development since the Rio conference in 1992. Aren't debates about the implications of a new geological era just extensions of all-too-familiar debates about known sustainability issues, aged solutions, and aged lines of conflict?

An interpretation of that sort, would in my mind, be fundamentally flawed. There are four aspects in this debate that illustrate how the human enterprise's move into 'planetary *terra incognita*' (Steffen et al. 2007) actually embeds some intricate and novel institutional and political challenges compared to earlier debates.

First of all, *nonlinearity*, *'catastrophic shifts'* or *'tipping points' matter*. The fact that complex social–ecological systems which underpin human well-being – such as coral reef ecosystems, agro-ecological landscapes, forests and freshwater – have the ability to shift rapidly and practically irreversibly to damaged states, add a whole different set of institutional challenges to global change. In short, 'tipping points' bring to light the need to explore the capacity of governance to facilitate early warnings, promote flexibility to changing circumstances, and support adaptive responses. This is an issue that will be explored in detail in the next chapter.

Second, *scale matters*. The nonlinear properties of vital social–ecological systems are not limited to local or regional scale examples. On the contrary, it has been suggested that these phenomena also characterize the behavior of the Earth system. Abrupt climate change (Alley et al. 2009), 'tipping elements' in the Earth system (Lenton et al. 2008), 'planetary boundaries' (Rockström et al. 2009b), a proposed possible 'state shift' in the Earth's biosphere (Barnosky et al. 2012), are all examples of attempts to explore the possibility of rapid, aggregated and destructive change up to the global scale. While these suggestions remain debated, the very prospect adds a layer of what I believe are unprecedented institutional requirements for governance at multiple scales.

With the future of planet Earth at stake, we should not be surprised that *politics matter* as well. As the previous discussion about 'planetary boundaries' clearly illustrates, defining a 'safe operating space for humanity' is bound to be a controversial and politically contested topic. As social actors such as non-governmental organizations (for example, *Oxfam-UK* and the multi-NGO *Planetary Boundaries Initiative*) mobilize to reframe their activities to match the framework, other societal interests (such as the *Breakthrough Institute*) counter-mobilize by challenging the underlying science, and its implicit policy implications. Sustainability scientists risk being caught in the crossfire as they attempt to offer tangible connections between science and policy (for example, Steffen et al. 2011b, Biermann et al. 2012, DeFries et al. 2012, Galaz et al. 2012b). While these controversies are far from new in global environmental politics (climate change being a prime example), the electric mix between possible global 'tipping points', science, and politics should not be underestimated.

Lastly, *technology matters*. Allow me to elaborate on this last point as technology is a central theme of this book.

Why Technology Matters

How do we understand Earth system 'tipping points'? Can these be measured, monitored and governed? And what novel governance mechanisms would be required to support human well-being within the 'boundaries' of the planet? These issues are – both from a scientific and political perspective – inseparable from perceptions of technological change. Simply put, technology has at least three functions in this emerging Anthropocene debate.[12]

First, technological advances have – at least in most parts of the world – driven increased human well-being at an impressive rate. Innovations such as antibiotics for medical uses; the Haber–Bosch process which creates millions of tons of nitrogen fertilizer for agriculture each year; and information and communication technologies, are just a sample of technologies which have been critical for the 'great acceleration' and humanity's move into a new geological era. At the same time, this acceleration has not only had unintended environmental and social consequences, but could potentially push the Earth system into an extremely turbulent future as environmental stresses propagate through teleconnections (Millennium Ecosystem Assessment 2005, Adger et al. 2009, see Raudsepp-Hearne et al. 2010 for an interesting synthesis of different positions in this debate).

Second, technological innovation has been critical for our current scientific understanding of the Earth system. Rapid advances in computing power in the few last decades have not only drastically facilitated our ability to monitor and model complex Earth system processes, such as the global climate system. They have also vastly expanded the geographical scope of scientific collaborations through the explosion of information and communications technologies. As shown by the Royal Society (2011), today over 35 percent of articles published in international journals are internationally collaborative, up from 25 percent 15 years ago. Information and communication technologies are also supporting the emergence of participatory science–citizen virtual collaborations through novel information communication platforms, often denoted as 'crowdsourcing' (Wiggins and Crowston 2011).

Economist Brian Arthur's point is well suited for this context: technology is not only the result of scientific discovery; technology also lays the ground for new scientific breakthroughs (Arthur 2009). The discovery of the Higgs boson particle in particle physics, and insights about extremely complex land–ocean–climate feedbacks in the Earth system, would be impossible without the impressive advancements in computing power driven by Moore's Law.

Third, technology is an important power, risk and cost distributer. By that I mean that technologies distribute risks and costs across time, space and social groups in ways that sometimes are difficult to predict in advance. They can be called 'quick-fixes' – technological solutions used to solve complex social–ecological problems – say, recurrent floods with higher levees. While higher levees might work initially, they also increase vulnerability and risks over time as illustrated by the devastating impacts on New Orleans by Hurricane Katrina in 2005 (Sterner et al. 2006). Distributional impacts of technologies can be profound in other ways, such as the redistribution and sometimes diffusion of power as the result of information technological change (Krasner 1991, Castells 2009). Hence technological distributions matter, and, as will be explored in this book, they pose tangible governance challenges in the Anthropocene.

Lastly, different (and often implicit) assumptions about the dynamics and direction of technological change affects how seriously different social actors perceive the risks of transgressing critical Earth system 'tipping points'. There is both a scientific and a political aspect to this. From a scientific point of view, the question is what a more explicit focus on technological change can offer to global change research agendas. DeFries and colleagues (2012) suggestion to reframe the 'planetary boundaries' as solution-oriented 'planetary opportunities', attempts to place human ingenuity and technological advances at the very center of global change research. Westley and colleagues (2011) instead suggest a stronger focus on the institutional context of innovation, and the need to explore mechanisms to steer innovation in ways that help the human enterprise to live within the 'capacity of the biosphere' (see also Leach et al. 2012, Olsson and Galaz 2012, Galaz 2012).

The political debate has a different focus, and the potential for catastrophic 'tipping points' adds an interesting dynamic (elaborated in Chapter 2). On the one hand, the potential for catastrophic shifts triggers debates about the need to rapidly develop and/or deploy technologies to put a brake on unwanted bio-geophysical changes at the global scale. Current suggestions to geoengineer the Artic to avoid devastating reinforcing climate feedbacks (such as proposed by the Artic Methan Emergency Group); to deploy large-scale protective shields for coral reefs (Rau et al. 2012); and proposals to ramp up the use of genetically modified crops and nuclear power (Lynas 2011) are all driven by concerns of the implication of global *nonlinear* change. Opponents, on the other hand, criticize claims about 'hard' bio-geophysical boundaries, arguing that human societies always have been able to overhaul seemingly non-negotiable physical boundaries through technological innovation. As the American ecologist Erle Ellis puts it:

> The 'planetary boundaries' hypothesis asserts that biophysical limits are the ultimate constraints on the human enterprise. Yet the evidence shows clearly that the human enterprise has continued to expand beyond natural limits for millennia. Indeed, the history of human civilization might be characterized as a history of transgressing natural limits and thriving. (Ellis 2012)

Diamandis and Kotler (2012) make a similar argument with an explicit focus on the potential of exponential development in technologies such as nanotechnology and synthetic biology, and additional changes in the global innovation landscape such as Do-It-Yourself-Innovators and billionaire techno-philanthropists such as Bill Gates and Richard Branson. In short, an analysis about governance and politics in the Anthropocene simply cannot afford to ignore the role of technological change.

CONCLUSION

The Anthropocene and associated scientific hypotheses such as 'planetary boundaries' are messy, socially contested, and scientifically debated. This, however, does not render these concepts useless to study from an institutional and governance perspective. Earth system science has always been debated. Geologists in the eighteenth century not only expanded our evidence-based understanding of the behavior of our planet, they also challenged deeply held religious beliefs about the Earth (Boardman 2010). In the beginning of the 1900s, the scientific suggestion that Earth was covered by slowly moving crustal plates (today known as plate tectonics), was widely challenged and debated by scholars and intellectuals.

This latter example is not intended as a harsh, populist argument against constructive criticism summarized earlier. Instead, it should be seen as a modest call to appreciate existing controversies about the Anthropocene and 'planetary boundaries' as a normal state of scientific discovery. Scientific theories evolve over time *because* they are debated, not the other way around.

My personal view in this debate, and in this book, is the following. While there certainly are areas of strong disagreement, few global change scholars would question the claim that humanity has entered a new era of rapid environmental change. The human enterprise has drastically changed Earth's climate dynamics, fundamentally modified marine and land-based ecosystems at very large scales, and even changed the course of biological evolution. These changes, in combination with insights about how complex systems behave, pose extremely challenging risks at unprecedented temporal and spatial scales. The rest of this book will build on this insight (or assumption, depending on where you stand in the debate).

The critical question therefore is not whether the Anthropocene from a stratigraphic perspective is here, or whether global 'boundaries' make sense at local and regional scales. Nor is it whether the proposed transgression of Earth system 'tipping points' could lead to global collapse, or instead indicate tremendous 'planetary opportunities'. The critical and interesting questions – at least for me as a political scientist – are what this new Anthropocene debate brings to existing political discussions about global change, sustainability and governance. And how to analytically engage with the novel institutional and political challenges that emerge as humanity increases its domination over a complex Earth system. This issue is discussed in the next chapter.

2. Governance and complexity

Discussions about environmental governance, especially at the international level, are increasingly stirred by claims of increased complexity. What is 'complexity' really? In its more daily use, 'complexity' represents a catch-all word that apprehends 'messiness' and 'entangledness' in general. For critics, claims about increased complexity are just another one of those trendy and empty buzzwords that have propagated in intellectual debates in the last decade (McKelvey 1999). Its progress is impressive and spans across areas as diverse as business leadership studies, health sciences, and organizational theory, to studies of the behavior of financial markets.

The approach in this book is different. My ambition is to build on a growing body of theoretical and empirical research on the features and behavior of *complex adaptive systems*. As I intend to elaborate in this chapter, emerging insights about the behavior of these systems in human–environmental or social–ecological settings, pose a number of intriguing challenges for institutional analysis and governance research.[13] In addition, notions about the risks posed by nonlinear change often denoted as 'tipping points' or 'threshold effects', are creating new political conflicts only recently explored by the scientific community. In this chapter I explore these two issues, and highlight some important gaps in our understanding. I conclude by presenting three 'governance puzzles' based on this and the previous chapter, and briefly introduce the case studies in the second part of the book.

COMPLEXITY AS THRESHOLDS, SURPRISES AND CASCADES

Complexity has for decades captured the interest of scientists and the public. One early and popular concept that has had a lasting impact on public perceptions is chaos theory, and the well-known 'butterfly effect' – the ability of small changes in one place to propagate and become amplified across unexpectedly large scales. And for those interested in pop-culture, there is also a Hollywood movie 'The Butterfly Effect' starring Ashton Kutcher with the slogan 'Such minor changes, such huge consequences'. But complexity theory – here explored through its subfield of *complex adaptive systems* – is a much richer theoretical field, and expands beyond unpredictable chaos and catchy taglines.

There is not one all-encompassing 'complexity theory', but rather a number of different research traditions, ranging from systems theory and cybernetics, to economics and human–environmental systems. These all pursue diverse methodological agendas, but share the theoretical assumption that a diverse set of real world phenomena in nature and society emerge as the result of complex interactions between agents and across scales. Compared to simpler (in other words, linear) systems, the behavior of complex adaptive systems is driven by *adaptive agents* (for example, cells, species, social actors, firms, and nations) who respond to locally available information. As agents attempt to adapt to changing circumstances, temporary and unstable system equilibriums emerge driven by processes denoted by *positive and negative feedbacks*. These feedbacks can either stabilize a system (for example, in the way a thermostat does by regulating temperature), or result in rapidly shifting system behavior, often with limited predictability.[14]

One simple example would be markets. As banks, firms, and investors (*agents*) try to break beneficial deals based on available market information, they not only create certain aggregate predictable patterns (for example, a certain price behavior defined by supply and demand); their actions can also create unexpected shifts such as market crashes as re-enforcing feedbacks kick in (Sornette 2004). This sort of unexpected behavior stemming from interactions in complex systems is also what is denoted as emergence, or emergent behavior.

Essentially, this is precisely the sort of mechanism that alarm Earth system scientists as they elaborate planetary processes that could trigger very rapid environmental change, such as Tim Lenton and colleagues' elaboration of 'switch and choke points' in the climate system (Lenton et al. 2008). It is important to note that these feedbacks can interplay both across human–environmental and between biophysical systems. In 1997 for example, Schellnhuber and colleagues suggested that many important global environmental problems, such as desertification and overexploitation of natural resources, could be portrayed as feedback processes between human and environmental systems, or what they denoted as 'syndromes of global change'.

Six and colleagues (2013) recently elaborated the ways in which climate change and ocean acidification interplay with marine organisms in ways that amplify global warming. In short, acidification might lead phytoplankton to emit less dimethylsulphide (DMS), a compound that helps to seed the formation of clouds. As this compound decreases, so does the formation of sun-reflecting clouds thereby amplifying warming.

A number of intriguing phenomena have been associated with complex systems behavior – such as 'chaotic change', 'hysteresis', 'strange attractors', 'bifurcation', and 'self-organized criticality'.[15] These phenomena differ in their underlying mechanisms and definitions, and are hard to harness using conventional governance or institutional analysis (Ostrom 2005, Duit and Galaz 2008).

Here I will simplify the analysis by focusing on the governance challenges posed by a limited selection of theoretically acknowledged and empirically well-elaborated categories of system effects: namely *thresholds*, *surprises* and *cascades*.[16] As several studies elaborate, these properties are general enough to capture interesting complex system behavior, as well as being applicable on a wide set of real-world phenomena (Duit and Galaz 2008, Galaz et al. 2008).

Thresholds

One key feature of complex adaptive systems is their ability to display 'threshold behavior', sometimes also denoted as 'tipping points', 'regime shifts', 'catastrophic shifts' or 'abrupt change'. The terminology differs, and the precise mathematical formalizations are diverse depending on the research field (Scheffer 2009). Nonlinear dynamics of this kind have been explored in the social sciences as well. Mark Granovetter's article from 1978 of threshold effects in collective action is a classic and early example of complex systems thinking for social phenomena. Paul Pierson's (2003) elaboration of how abrupt social change can be triggered by minor disturbances in political systems, and Stephen Krasner's notion of punctuated equilibrium, provide additional examples (Krasner 1984) and are two parallel explorations in the same tradition (see Duit and Galaz 2008 for references and synthesis).

The common theme is that small events might trigger abrupt system changes that are difficult, or even impossible to reverse. In some cases the transition is sharp and dramatic. In others, the transition itself may be slow, but definite. Hence, seemingly stable systems such as coral reef ecosystems can suddenly undergo comprehensive transformations into something entirely new (for example, from coral to algal dominance), with internal controls (in other words, feedbacks) and characteristics that are profoundly different from those of the original (Gunderson and Holling 2002, Kinzig et al. 2006). Threshold effects of this sort have attracted wide interest and empirical validation for a number of real-world systems, including physics (Goldenfeld and Kadanoff 1999), ecological systems such as shallow lakes, and coral reef ecosystems (for example, Folke et al. 2004, Scheffer and Carpenter 2003), and even key Earth system functions such as the climate system (Lenton et al. 2008).

It should be noted that the use and analysis of thresholds in the academic world is diverse. Thresholds can differ in their temporal dimensions, their degree of reversibility, and the number of possible alternative states, just to mention a few dimensions (see Scheffer 2009 for an elaboration). In addition, thresholds can also interact at different scales (from local to global) in very complex, poorly understood (Kinzig et al. 2006) and contested ways (Brook et al. 2013). As will be explored below, the suggestion that thresholds have been proposed to exist at the global scale, captured in the suggestion of multiple interacting 'planetary boundaries', has induced intense academic and political debates (elaborated in the previous and the next chapter).[17]

Interconnectedness and Cascades

Another key property of complex adaptive systems is their interconnectedness, which is the ability of components to be linked to one another. For example, a complex technology such as the World Wide Web, or financial markets, evolve and change as the result of multiple interacting components (hardware, software and computer users). In a similar way, ecological systems and the Earth system change and respond to changes as connected physical, chemical, biological, and human components influence each other. The tight interconnection between humans and the biosphere is often captured under the umbrella terms 'human–environmental systems' (Schellnhuber et al. 1997), 'coupled human and natural systems' (Liu et al. 2007) or 'social–ecological systems' (Berkes et al. 1998). This definitional choice might seem obvious, but is not trivial as it forces scholars to study the dual relationship between social systems – including human behavior, institutions, policies – and changes on ecological systems and the biosphere (Folke et al. 2005). As a political scientist, I know for certain that this sort of approach is far from common in my discipline.

Interconnectedness also implies the disposition of complex adaptive systems to be subject to cascading disturbances. Liu and colleagues (2013) explore the underlying features of interconnectedness defined by 'flows' creating what they call 'telecoupling' dynamics. 'Flows' are movements of material, energy or information between systems (human, and/or environmental).

Of interest for the purpose of this book is the observation that increased interconnectedness also implies that disturbances can unfold *across scales* (for example, from local to regional to global), *in time* (for example, delayed impacts), and/or *through sectors* (for example, from the technical to the economic or political system) (see Helbing 2013). For example, the transgression of 'thresholds' for coral reef ecosystems due to ocean acidification is not only an ecological concern; it also raises concern from a human well-being perspective due to its likely impacts on marine species, and associated food security problems for coastal communities.

The likelihood of cascades is related to the degree of coupling between systems, an aspect often called 'modularity'. The argument is that loosely coupled systems have more time to recover from failure and are therefore better able to buffer potential cascades, while tightly coupled systems do not allow time for delays and thereby increase the risk that disturbances become amplified (Perrow 1984). There is a difficult trade-off here: too disconnected systems (such as coral reefs and forests) are more vulnerable as their isolation undermines sources of recovery, for example the transportation of nutrients, larvae and juveniles from surrounding areas (Nyström and Folke 2001).

Surprises

Thresholds and connectivity are two important complex systems phenomena. A third feature of interest is surprise – events that occur when system behavior differs qualitatively from a priori expectations. The possibility of 'surprises' such as terrorist attacks, earthquakes, riots and similar shocks has been a central theme in crisis management research (Boin et al. 2005). Some of these have been framed as 'low probability, high impact' events, or 'Black Swans' as popularized by Nassim Taleb (2007). 'Surprises' are also central in resilience thinking (Gunderson and Holling 2002). The underlying reason for surprise stems from the fact that complex systems contain poorly understood interactions driven by both positive and negative feedback, and processes operating over a range of spatial (local to global) and temporal (fast to slow) scales. This might sound unnecessarily over-theorized, but consider the underlying combination of factors that drove the 2008 'food crisis' which led to rapid increases in food prices globally. As I will explore later, the combination of phenomena such as droughts, increasing oil prices, shifts in investments patterns amongst financial actors, shifting diets, and the expansion of biofuels, added up and recombined in ways that seriously challenged the coping capacities of national governments and international organizations. Global shifts, national responses, human and environmental factors all interplayed through telecoupling in ways never experienced before. While the underlying causes are still being debated, studies clearly show that the events caught policy-makers by surprise as they attempted to make sense, and manage an unfolding yet poorly understood crisis (Galaz et al. 2010b).

Interlinked Features in the Real World

These features might sound like highly abstract theoretical concepts. In fact, a number of intriguing phenomena unfolding in an age of increased interconnectedness and rapid environmental change benefit considerably by being analysed from a complexity perspective. In addition, a closer look at the dynamics of real-world sustainability challenges is highly illustrative of how surprise, thresholds and cascades interplay in complex ways. A number of scholars have raised this point before (Young et al. 2006, Kinzig et al. 2006, Adger et al. 2009, Galaz et al. 2010b), but the argument is worth some elaboration.

Epidemic outbreaks triggered by zoonotic diseases (in other words, animal diseases that spread to humans) are an interesting case in point (to be explored in detail in Chapter 4). While it has been suspected for considerable time that a potential devastating animal flu might evolve rapidly, infect humans and move across geographical boundaries, scholars have constantly struggled to make sensible risk assessments of where and when this is most likely to happen. As scholars increasingly focused on Asia and Africa as epidemic 'hot-spots', the 2009 pandemic flu (A/H1N1 or 'swine flu') unexpectedly had its origin in North America. This surprise triggered a number of responses as decision-makers attempted to stay beyond critical epidemic *thresholds* (as a transgression would imply infection rates beyond control), as well as to mitigate potentially serious *cascading* economic and social impacts.

Homer-Dixon (2002, p.173) explores the notion of three coupled systems – natural, social and technological. As this book will show, I very much share his argument that 'we have produced incremental quantitative changes in all three types of system. By themselves, the changes have usually been tiny in degree or amount, but together they have produced shifts in the quality of these systems' basic character and behavior'. Not all real-world phenomena display all of these three properties of course, but their interactions and associated governance challenges should not be underestimated (Walker et al. 2009).

INSTITUTIONAL DIMENSIONS OF 'TIPPING POINTS' AND THRESHOLDS

The ability of social–ecological systems and Earth system functions to change abruptly, and to practically irreversible states has attracted considerable interest from the scientific community in the last few decades (Scheffer et al. 2012, Rockström et al. 2009a, b, Lenton et al. 2008).

However, while natural scientists are becoming increasingly capable at defining, measuring and proposing 'tipping points' in bio-geophysical systems at several scales (for example, ranging from shallow lakes or Earth system function), the new political tensions these unleash have been considerably less investigated. These heated discussions about 'tipping points' bring to light poorly explored governance challenges in the Anthropocene. This section elaborates these tensions in detail, identifies major gaps in our understanding from a governance perspective, and suggests some way to bridge them.

The Contested Nature of 'Thresholds'

Efforts to avoid the transgression of 'tipping points' might seem like a rather straightforward challenge for decision-makers. A similar management model to that applied to toxic chemicals could in principle be applied – define a scientifically based threshold value (say, x for chemical or pollutant y), add a safety margin, and implement policies that keep you within the defined safety margin, or phase its use out entirely. As the scientific evidence changes, both the threshold value and the safety margin can be adjusted.

While the approach might seem oversimplistic, the logic is similar to political attempts in climate policy to define a 2°C target,[18] as well as the approach laid out by Rockström and colleagues (2009b) in their definition of planetary boundaries demarcated by 'scientifically informed values of the control variable established by societies at a "safe" distance from dangerous thresholds'. In addition and at best, long time data series and sophisticated statistical methods would allow scientists to extract early warning signals of a pending transgression of a threshold timely enough for social responses to be able to steer away from collapse (see discussion below).

As Mike Hulme notes, 'thresholds' are more than scientifically defined and straightforward numerical estimates. Threshold targets also contribute to existing narratives about environmental urgency, and the risk of large-scale collapse unless swift action against environmental destruction is taken (Hulme 2012). Biermann (2012) and myself in Galaz et al. (2012a) expand the argument in explorations of how 'planetary boundaries' could support a coherent reform of global environmental governance in the Anthropocene. Hence scientific explorations of 'tipping points' are subtly yet consistently linked with implicit assumptions about the need for preemptive social and political response (Young 2012).

Trying to build momentum for political action based on pending transgressions of critical environmental thresholds, has been contested, to say the least. Mike Hulme's critical reflection of the 2°C climate policy target (Hulme in Knopf et al. 2012) argues that a simple numerical target not only hides critical scientific uncertainties (see also Tol 2007), but also endangers pushing other critical global goals such as the alleviation of poverty to the background. By 'being abstract, distant in time and ambiguous', thresholds obscure other critical issues and 'allow[s] politicians to evade its demands as to encourage them to embrace them' (p. 124).

Interestingly enough, these are similar to the arguments advanced by critics of 'planetary boundaries'. Lewis (2012) for example, argues that not all of the 'boundaries' embed the sort of global threshold dynamics that lead to abrupt Earth system change (using phosphorous as an example). And again: the definition of global 'boundaries could spread political will thinly – and it is already weak' (Lewis 2012). Blomqvist and colleagues (at the *Breakthrough Institute*, mentioned in Chapter 1), question the scientific basis for the quantification of 'planetary boundaries', and worry that they 'may risk misleading local and regional policy choices' (2012, p. 6) and that proposed boundaries 'may harm the policy process, as it precludes democratic and transparent resolution of these debates, and limits, rather than expands, the range of available choices' (2012, p. 8). Anthropologist Mark Nuttall suggests that a focus on 'tipping points' implies a simplistic and deterministic discussion about the future, and raises 'fears of uncertainty, danger, risk, disasters, and catastrophe' (Nuttall 2012, p. 99).

Hence a scientific and social debate about the fruitfulness of defining, measuring and trying to govern global environmental challenges entrenched by 'thresholds' or 'tipping points', is rapidly unfolding. It should be noted that these increasingly loud discussions seldom are based on a solid *empirical* understanding of how decision-makers perceive and respond to systems that embed 'tipping point' properties. In my opinion, any confident claim that scientifically defined Earth systems or ecological 'thresholds' undermine effective policy-making (Blomqvist et al. 2012), divert away attention from other urgent issues (Hulme in Knopf et al. 2012), or spread political will 'thinly' (Lewis 2012), need to be backed up by evidence. This could entail case studies, modeling or any sort of empirical and theoretical analysis of the processes which lead decision-makers to fail (or succeed) in trying to respond to threshold dynamics. At present, they are not.

Understanding Social Perceptions and Responses to 'Thresholds'

Luckily, several research streams attempt to systematically analyse how social actors, such as policy-makers and natural resource users, respond to the risks of transgressing 'tipping points'. These streams (illustrated

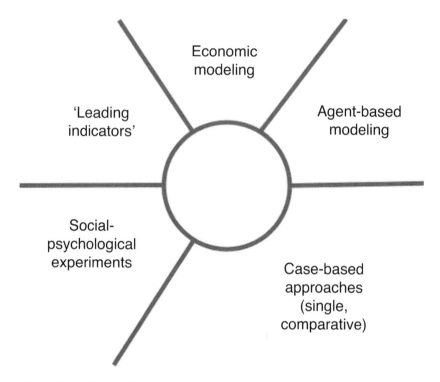

Figure 2.1   Research approaches to 'tipping points' and their social responses

in Figure 2.1) range from pure theoretical analysis in economics, and interdisciplinary systems modeling approaches, to social-psychological experiments and case-based empirical studies. Methodologies, theoretical assumptions, and data gathering techniques differ considerably between these fields of course. In addition, several attempts have been made to combine some of these methods, which make a stringent separation difficult. Taken as a whole, they give a far more nuanced picture of the governance challenges posed by 'tipping point' change in complex systems (Luke and Stamatakis 2012).[19]

Economic Modeling

One of the earliest theoretical attempts to analyse social responses to threshold behavior emerged within neoclassical economics. While the body of research is large, it is clear that threshold dynamics pose quite intricate decision-making challenges. The earliest results from the mid-

1970s indicated that breaching a threshold by depleting a natural resource could be an optimal economic strategy. In addition, if a threshold is known with certainty, optimal policies would let degradation or resource extraction unfold gradually, up to the threshold level (see review by Chen et al. 2012). Hence knowledge about possible devastating abrupt changes if a system is pushed too far would, according to this reasoning, not slow down degradation but rather accelerate it, up to a certain optimal point.

Attempts three decades later, however, increasingly integrate insights from ecological studies, including an acknowledgement of the possibility of multiple equilibrium points, and more realistic assumptions about threshold uncertainty (in other words, uncertainty about the precise location of the 'tipping point' see Carpenter et al. 1999, Mäler et al. 2003, Crépin 2007). These studies instead highlight that conventional optimization policies are not only difficult to implement in systems with threshold properties, but are also dangerously counterproductive. In essence, as decision-makers only superficially understand complex systems, they tend to overestimate the carrying capacity and resilience of their system. As policies are implemented based on these simplified assumptions, the modeled 'social planners' unavoidably push systems beyond critical thresholds, eventually inducing abrupt unwanted environmental changes, for example fish stock collapse, practically irreversibly degraded shallow lakes, or ruined coral reefs. Optimal management strategies for systems with threshold are difficult to identify, as they would require full knowledge about system behavior. Ironically, this would require actors to intentionally push the system of interest beyond the threshold they originally wanted to avoid (Crépin 2007, p. 208). As Anderies and colleagues (2006) summarize it: 'The main insight is that more cautious policies are required to avoid welfare losses when there is the possibility of moving from one regime to another. The level of caution increases when managers are uncertain about the location of thresholds or there are time lags in policy implementation' (p. 867).

Agent-based Modeling

Computer modeling of complex and coupled social and ecological systems – including both systems and agent-based modeling – have also explored how feedbacks and 'tipping points' are shaped by human responses. Schlüter and colleagues' synthesis (2012) bring together a rich literature with varying theoretical assumptions about human behavior, system dynamics and levels of interplay between social and ecological factors. Early simpler modeling attempts in the 1990s attempted to elaborate optimal strategies, for example for rangelands which are prone to 'tipping point' vegetation changes, or for example from grassland to a shrub-dominated land. These models have become increasingly complex over time and include both ecosystems with nonlinear behavior and feedbacks, adaptive agents with the ability to learn from changing circumstances, as well as governance structures (for example, Heckbert et al. 2010, Horan et al. 2011).

General conclusions from this field are difficult to draw, due to their diversity in focus, underlying theoretical assumptions and conclusions. Three studies stand out as illustrations of subfields that explore the complex link between 'tipping point' behavior and human action.

Carpenter and Brock (2004) show how local collapse of fish populations can lead to cascading collapse in neighboring fisheries as fisherman shift between locations. Janssen and colleagues (2004) use a genetic algorithm to explore the possibility for robust strategies for rangelands, that is, management options that are able to maximize expected returns despite big uncertainties and the possibility of thresholds.

Bodin and Norberg (2005) instead use computerized agents – that is, an actor acting under pre-specified rules created by computer software – to understand how different patterns of information sharing ('information network typologies') affect the extraction from a resource with threshold properties. One important conclusion is that strongly connected actors not only were able to share information about successful harvesting strategies, but also that this information led to cascading over-harvesting and large-scale collapses. As communication links become less dense, harvesting levels vary and are lower, but are more robust as they provide actors' with the ability to learn and recuperate from their neighbors' mistakes.

Axtell and colleagues (2002) instead combine agent-based modeling with archeological records to understand how human–environmental interactions, such as demographical change and resource extraction, led to the rapid growth and collapse of the *Anasazi*, a culturally rich empire inhabiting northeastern Arizona 1800 BC to 1300 AD. As others similar pre-historical studies also explore, the tendency of societies to focus on simplification and efficiency, also tends to increase vulnerability to stresses and shocks (Schoon and Fabricius 2011). In essence, as societies invest in social and physical infrastructure, they find themselves increasingly 'locked in' and prone to abrupt nonlinear changes in both human and natural systems (Tainter 1988). For example, pre-historic societies benefited greatly from the creation of large-scale irrigation systems, an important technological investment that induces population growth and migration into cities. As changing climatic conditions and novel disturbances (droughts, floods) emerge, these create unprecedented shocks in societies where efficiency also has led to loss of alternative production strategies, rapid extraction of natural resources and a reduced capacity to migrate (see Nelson et al. 2010 for details).

Early Warning Signals ('Leading Indicators')

An important subfield to modeling focuses on the possibility of extracting early warning signals from large sets of data, using advanced statistical methods. The underlying assumption here is that complex systems in, for example, finance, ecology, engineering and microbiology, display similar mathematical properties and therefore can be fruitfully compared (Scheffer et al. 2012). For example, mathematical and experimental studies have shown that measurable key parameters in systems (for example, concentration of algae in shallow lakes) are prone to 'wobble' or 'stutter' before crossing a 'tipping point', sometimes 10–13 years before the transgression (Wang et al. 2012). Other early warning signals of a pending transgression include the disposition of complex systems to recover unusually slowly after small disturbances, or the ability of ecosystems such as savannah landscapes to show intriguing patterns of vegetation patchiness before a transition to a desert (see Dakos et al. 2012, Perretti and Munch 2012, Scheffer et al. 2009 for details).

While research on ecological early warnings is clearly on the rise, few have explored whether it is possible to extract reliable warnings early enough for societal responses to be effective. *Reliability* and *timeliness* are critical issues from a governance perspective, especially considering that complex systems entail irreducible uncertainties (which imply noisy data, uncertainty about whether there is a threshold at all, and the risk of false early warnings) as well as considerable time lags between action and effect. Biggs and colleagues (2009) and Perretti and Munch (2012) are two important exceptions. As Biggs and colleagues (2009) show for a fisheries food web model, clear indicators of a pending fish stock collapse only emerge when it is too late to reverse the unfolding transgression. Perretti and Munch (2012) also raises the point in their simulations, that while useful early warnings might be possible for simpler closed systems, larger and open ecosystems such as marine ecosystems seriously increase 'noise' in data, and make early warning signals practically impossible to differ from normal variation (see also Groffman et al. 2006, Boettinger and Hastings 2012).

More causally phrased: as you drive a fast car on a highway in the evening, several scenarios are possible. One is that you suddenly get a clear view of an unexpected approaching cliff edge, but not timely enough for you to be able to steer away. Another is that the edge is impossible to distinguish from a chaotic landscape dominated by digital billboards, obfuscating buildings, and an ever-changing panorama. Or lastly, that your foot on the gas pedal not only forces the car to accelerate, but also simultaneously pushes the edge closer to you.

Experimental Approaches

Another related approach instead builds on laboratory experiments. The application of experiments where subjects (often university students) under different experimental circumstances are allowed to extract from a common pool resource, has a long tradition in studies of common dilemmas (such as 'tragedy of the commons' and 'public good dilemmas'). Laboratory experiments have been critical for unpacking the role of group size, heterogeneity and communication for collective action (Ostrom 1998). The experimental settings have also become increasingly sophisticated over time including attempts to explore spatial and temporal dynamics (Janssen et al. 2010), computer simulations (for example, Jager et al. 2002) and by involving real-world natural resource users in the experiments (for example, Cárdenas and Ostrom 2004).

The behavior of subjects in response to an imminent dangerous threshold is a fairly recent research area, but has already generated some interesting results. Therese Lindahl and colleagues (2012) for example, design an experiment that is able to compare two treatments. In both cases, users face a common pool dilemma, but while some groups face a dilemma with simple (linear) resource dynamics, other groups face complex nonlinear resource dynamics involving a potential known threshold change in the resource. The results indicate that groups that face a resource with potential threshold behavior 'are much more likely to cooperate and obtain efficient harvest levels [. . .]. As a result these groups are also less likely to over-exploit and deplete the resource' (p. 25). The mechanism proposed is that actors perceive thresholds as a common threat, which consequentially acts as a catalyst for communication, trust building and eventually cooperation (see also Milinski et al. 2008, Barrett 2011).

In other experiments, Barrett and Dannenberg (2012) show that collaboration of this sort is largely dependent on whether subjects have knowledge about the location of the threshold. That is, their experiments show that certainty about the location of a threshold, transforms commons games into one of coordination. In this case, simple communication between the subjects is enough to create collaborative outcomes. However, the results also show that as soon as there is some uncertainty about the location of the threshold, collaboration tends to collapse. The mechanism here is that while actors have a collective incentive to avoid the far-reaching consequences of exceeding a threshold, they also 'face individual incentives to free ride because of the inherent uncertainty about the location of the threshold' (Barrett and Dannenberg 2012, p. 17372). Applied on climate change negotiations, their research suggests that 'countries are very likely to propose to do less collectively than is needed to avert catastrophe, pledge to contribute less than their fair share of the amount proposed, and end up contributing even less than their pledge'.

Case-based Approaches

The ability of ecosystems to shift rapidly to a different stable state due to threshold dynamics has also attracted wide interest by scholars of ecosystem governance. Common umbrella terms for this research field are 'adaptive co-management' or 'adaptive governance', and aims to explore the features of governance modes able to adjust and learn from changing circumstances, and cope with stresses and shocks. Empirical support is often based on local or regional studies (individual or comparative case studies) including the collection of primary data from interviews, document studies and on-site ecological studies (see synthesis in Plummer et al. 2012).[20]

Adaptive co-management and governance approaches are often presented as an alternative to centralized and hierarchic governance arrangements, sometimes denoted as the 'pathology of natural resource management' (Holling and Meffe 1996), or 'panaceas' (Ostrom 2007). Hence the ambition is not only to explore the challenges of 'tipping points', but instead the capacities of actors and institutions to cope with multiple forms of shocks and stresses. This field is quite diverse, which brings certain empirical and theoretical richness into the academic debate, but also implies a lack of a general overarching and unifying analytical framework (Plummer et al. 2012, Ostrom 2007).

Two attempts have been made to summarize, and bring coherence into the field. In 2003, Dietz and colleagues (2003) laid out the underlying governance structures and processes that support adaptability and flexibility. These features include the capacity to provide fine-grained information across scales; the capacity of institutions to resolve conflict and induce compliance; as well as the ability of users to provide necessary technological infrastructure and adapt to changing environmental circumstances.

The lengthier synthesis of the field by Carl Folke and colleagues (2005) identified similar features, but placed more emphasis on issues related to the combination of different forms of knowledge (for example, scientific expertise and local practitioners' ecological experience), social network structures across levels, and high degrees of social capital and memory. Despite different approaches on the issue, both syntheses highlight *trust, learning, institutional flexibility, diversity and multi-level linkages* as factors that support successful attempts to govern complex social–ecological systems.

A closer look into individual studies of how actors explicitly respond to 'tipping point' dynamics in ecosystems can nevertheless be found in studies such as Olsson et al. (2006), Gelcich et al. (2010), Anderies et al. (2006), and Young (2012). Again, drawing simple general lessons on how natural resource users and decision-makers respond to 'tipping point' behavior is difficult, and results differ depending on the theoretical approach. One stream of research indicates that the threat of a threshold transgression has the ability to create a sense of urgency, and help mobilize actors in ways that unbolts institutional lock-ins. This seems applicable at both regional scales (for example, municipality level, Olsson et al. 2006), at national (Gelcich et al. 2010), and the transnational (Olsson et al. 2008) and international level (Österblom and Sumaila 2011).

Another stream of empirical studies instead highlights how irreducible uncertainty and institutional path-dependence drive decision-makers to push their system across thresholds. Using the Goulburn Broken Catchment (Australia) as an example, Anderies and colleagues (2006) combine modeling and historical institutional analysis to track how decision-makers since the nineteenth century have responded to ecological crisis (for example, surprisingly rapid and large-scale raising water tables) by centralizing water governance structures, implementing engineering solutions, and in the longer term increasing the risk for large-scale abrupt changes in multiple biophysical systems such as the depth of the water table, biodiversity, and soil acidity (Walker et al. 2009b, see also Allison and Hobbs 2004).

Summary

Thresholds dynamics entail a number of challenges for the possibility to design institutions nimble enough to detect, prevent and respond in a timely fashion. As I've elaborated above, a rich and multidisciplinary field exploring the decision-making and governance challenges posed by 'tipping points', does indeed exist. Results are difficult to summarize, and differ considerably depending on theoretical approach, methods and main focus. Table 2.1 is an attempt to summarize these diverse research streams.

Some evidence points at the serious difficulties involved in trying to abstract 'early warnings' of pending thresholds transgressions from data, how social actors are likely to fail to coordinate in the face of uncertainty, and how decision-makers are bound to transgress thresholds as a result of path-dependent institutions and technologies. Other studies indicate the potential of novel statistical approaches to extract to 'early warnings', how

Table 2.1 Literature synthesis

	Emphasis	Actors, Scale and System	Main Conclusions
Economic modeling	Formal theoretical approaches to explore optimizing challenges posed by natural resources with nonlinear properties.	Hypothetical decision-maker or 'social planner', local resource, for example, lake or coral reef but conclusions intended to be general.	Breaching of threshold or exploitation up to a known threshold value can be optimal. Later work shows that threshold uncertainties result in counterproductive policies as 'social planners' unavoidably push systems across thresholds, leading to collapse of resource.
Agent-based modeling	Computer modeling of agents and their interplay with environmental factors. Explicit focus on system understanding, feedbacks, social–ecological interplay.	Depends on specific model, often local users managing a local or regional resource such as water or land.	Inconclusive, depends on model assumptions. Key studies explore existence of robust strategies; combinations with real-world data explore dynamics of human–environmental collapse; modeling and social network analysis highlight risks of cascading collapse due to over-connection.
Leading indicators	Extraction of 'early-warning signals' or leading indicators of approaching 'tipping points' through statistical analysis of long time-series.	Social actors not part of models, emphasis on complex systems behavior in general. Combinations of mathematical modeling and empirical data analysis for lakes (local and regional), and the climate system (global).	Evidence of generic leading indicators or 'early-warning' signals in several complex systems before a pending transgression of threshold. Emerging discussion on practical usefulness due to challenges in reliability and timeliness.

Table 2.1 (continued)

	Emphasis	Actors, Scale and System	Main Conclusions
Social-psychological experiments	Focus on how subjects in experimental settings respond to collective action problems, including extraction of resources with diverse 'tipping point' dynamics.	Subjects (often university students) intended to represent generic collective action challenges. Extraction from an idealized open access resource.	Early results show that groups that face a resource with threshold are more likely to cooperate and obtain efficient harvest levels. However, threshold uncertainty has been shown to cause free riding and resource collapse.
Case studies (single and comparative)	Miscellaneous literature with attempts to identify features of adaptive co-management and governance, focus on multilevel institutions, knowledge generation, actor networks.	Local communities, indigenous groups, practitioners and wider set of natural resource users with special emphasis on interplay between social and ecological systems. Scale often local–subnational, but recent work also explore cross-national governance.	Studies show that the threat of a transgression of a threshold has the ability to unbolt institutional lock-ins and helps mobilize actors and institutional change. Other studies emphasize difficulties of avoiding thresholds as decision-makers are unable to deal with institutional and technological path-dependence, and irreducible uncertainty.

institutional entrepreneurs successfully can tap into a sense of urgency to promote institutional change, and with proper information promote collective action before a threshold transgression.

While these mixed results may not come as a surprise, it is interesting to note how they challenge and provide a more evidence-based picture of how social actors – ranging from natural resource users to decision-makers at the policy level – perceive and respond to 'tipping points'. Hence confident claims that scientifically 'thresholds' undermine effective policy-making or spread political will 'thinly' are premature at best, and misleading at worst. The important question now is how to integrate these insights into a broader governance research agenda. This is the issue of the next section.

CRITICAL GAPS AND MOVING AHEAD

Can we scale up insights from studies of successful governance of complex social–ecological systems at local and regional scales to the global scale? This might seem like an overambitious task, but one that soon needs to be assumed by governance scholars. The reasons are quite straightforward.

First, current discussions about the Anthropocene indicate quite clearly that environmental stresses now play out in nonlinear ways (including 'tipping points') at the global scale (Rockström et al. 2009a, 2009b, Adger et al. 2009, Walker et al. 2009, Steffen et al. 2011). These combined human–environmental stresses pose an array of new challenging questions for the study of governance as it forces scholars to consider the ability of global institutional arrangements to address complex Earth system processes. While this might seem obvious, surprisingly few attempts have been made to explore the features of adaptive governance in a global setting characterized by connectivity across scales, complex social–ecological interactions, and nonlinear change. As related fields of research have proven, insights from local scale studies do not scale up to the global scale easily (Ostrom et al. 1999).

Second, the most intense political discussions that evolve at the very heart of the Anthropocene debate are essentially about the role of institutions, politics and governance at the global level. Are 'planetary boundaries' a useful approach to frame global policies such as the Sustainable Development Goals? Is geoengineering a useful approach to steer away from regional and global 'tipping points'? What scientific assessment processes are needed to monitor rapid environmental change with nonlinear properties? These are just three examples of important questions at the science–policy interface, the answer to which will require a more fine-grained understanding of the role of international actors, institutions and governance and their attempts to address nonlinear change at the global scale.[21]

One way to think about these challenges, is in terms of 'governance puzzles'; a number of generic, challenging and integrated governance questions with the ability to bring to light novel issues to explore as we move into a new geological epoch. Before defining these 'puzzles' more closely, I would like to just briefly return to some of the issues I brought up in the first chapter. These were related to my claim that four subjects make current discussions about governance in the Anthropocene different from known debates about sustainability: nonlinearity, scale, politics and technology. What's interesting here is not only their respective imprint on current discussions, but their interplay as discussions about nonlinear global change are entangled with politics and views on the potential of technological change. I believe three questions or 'governance puzzles' are critical in this new context.

Governance 'Puzzles'

The first is related to the international system's ability to deal with 'global human–environmental surprise'. By this I mean situations in which the behavior in a system, or across systems, differs qualitatively from expectations. While some of the impacts of global environmental change (such as climate change) can be predicted or at least estimated through modeling and scenarios, other events will unfold as surprise events. Recent examples here include the 2008 'food crisis' and its repercussions on food security and ecosystem services; outbreaks of novel zoonotic diseases; or extreme weather events that trigger unexpected social turbulence and political instability (Galaz et al. 2010b). Many of these surprise events can unfold within the coping capacity of institutions. Others, however, can propagate and create severe threats to human well-being. This is particularly the case when events do not match previous experiences; embed scientifically and socially contested cause and effect relations; and when information integration is challenged by organizational silos, and gaps in ecological monitoring (Galaz et al. 2010b). Hence the first 'puzzle' I would like to explore in this book is:

Governance Puzzle (1). What characterizes international institutions that are able to detect and respond to 'global human–environmental surprises' of large importance to human well-being?

The second 'puzzle' is not related to crisis events, but rather about the entangled underlying drivers of global change. The fact that several

Earth system processes (say climate change, biodiversity and land use change) interact in ways not fully understood by science, poses difficult governance challenges, especially as these are embedded in a complex web of institutions, information and technology (Walker et al. 2009, Adger et al. 2009). This requires us (political scientists and social scientists in general) to find ways to analyse the implications of a highly complex and nested Earth system for institutional analysis. The second 'puzzle' is therefore:

Governance Puzzle (2). Are international institutions at all able to address complex Earth system interactions, or should we instead put our faith on the emergence of polycentric approaches?

The third issue relates to the role of governance in steering innovation and technologies. Calls for large-scale support of technological innovation as a strategy to stay within critical Earth system thresholds (normally climate) are increasingly common in the debate. Proposals to cool down the planet through artificial volcanoes, or genetically designed algae with the ability to effectively convert greenhouse gases to fuels are, however, only a smaller aspect of a larger and rapidly changing technological landscape. Emerging technologies such as synthetic biology, communication and information technologies, nanotechnology and geoengineering without doubt pose intriguing opportunities to improve human well-being. At the same time these technologies also pose difficult and unprecedented challenges to governance as they also entail potentially large ecological risks. Creating the proper balance between experimentation, and precaution in settings where the transgression of thresholds is at play, is in no way a trivial question. Considering the rapid rate of technological change, it is likely to become increasingly important over time.

Governance Puzzle (3). Is a governance setting possible that is strong enough to 'weed out' technologies that carry considerable ecological risk, but still allows for novelty, fail-safe experimentation and continuous learning?

These three 'puzzles' will help guide parts of the analysis in the following case-based chapters, but more importantly, will summarize the analysis of the book to a wider non-academic audience. Without doubt many more questions and perspectives could be elaborated in a book about technology and governance in the Anthropocene. Hence, these should be viewed as a mere start of what is likely to become a rich field of scientific inquiry in the next decades.

THE CASES

The analysis assumed here is case-based. That is, I believe that in-depth case studies are essential to advance our understanding in such a complex issue area. While quantitative analyses certainly have an advantage through their ability to use statistical methods to generalize to a larger population, the theoretical field I explore here entails so many variables and is too theoretically immature to allow for such an analysis. In addition, my goal is explicitly to explore contested and poorly understood phenomena at the human–environmental–technological interface with some detail.

The next part of this book starts with the first case about *Earth System Complexity* (Chapter 3). In this chapter, I move beyond current political debates about 'planetary boundaries' and elaborate key international governance challenges posed by Earth system complexity, and some constructive ways to analyse these from a governance perspective. The chapter also includes an elaboration and discussion of global organizational networks and polycentric coordination in the face of complex human–environmental change.

The second case presented in Chapter 4, instead focuses on the complex institutional and governance challenges posed by emerging infectious diseases (EIDs) such as animal influenzas (for example, 'avian influenza' and 'swine flu'), and hemorrhagic fevers such as Lassa fever and Henipavirus. In this chapter I explore how international actors such as the World Health Organization try to stay ahead of epidemic surprise in terms of early warning and response. Here I also explore the role information and communication technologies play in the way international actors collaborate across cross-national networks, and how these networks interact with more formal institutions such as the Internal Health Regulations.

Suggestions of large-scale technological interventions to combat climate change that a decade ago would have been discarded as science fiction are slowly moving toward the center of international climate change discussions, science, and politics. Chapter 5 analyses the role of governance and complexity in the Anthropocene by taking a closer look at geoengineering technologies, and their contested international regulation. Here I place special emphasis on the governance tensions decision-makers face as they try to weed out geoengineering proposals that carry considerable ecological and social risks, but still allow for novelty, fail-safe experimentation, and continuous learning.

In the last case study chapter (Chapter 6), I analyse another emerging technology with implications for our ability to govern global change in the Anthropocene: algorithmic trade in commodity markets. Algorithmic trading (sometimes denoted as 'automatized trade', 'high frequency trade', 'computer based trading' or 'robot trade') is having profound impacts in the way and speed in which financial assets are traded. The capacities of computer algorithms to process increasing amounts of market information including financial news items, and conduce extremely rapid and complex trading patterns are clearly on the rise. The rapid advancement of algorithmic trade pose until now unexplored environmental governance challenges due to the increased connectivity between financial markets, commodity markets, and ecosystem services on the ground.

I'm very aware that the cases might seem quite dissimilar. But as I intend to show in the last chapter, they provide a rich and complementary picture of the intriguing governance questions that emerge at the interface between politics, nonlinear global environmental change, and technological change.

3. Earth system complexity

Have you ever heard about how deforestation in the Amazon affects precipitation patterns in Asia? Or how decreasing fish stocks in the coast of Central West of Africa triggers deforestation and the loss of wildlife in nearby African countries? If not, are you aware of how emissions of carbon dioxide not only result in climate change, but also induce ocean acidification, with possible associated losses of biodiversity, and changes in water cycles on regional scales?

These are all examples of how changes in the Earth system are interconnected in ways that are archetypal for the behavior of complex adaptive systems. These interconnections are both biophysical and social – that is, they are the result of flows of energy, material, species either through biophysical processes or as a result of human connectivity, including global trade, transport or communication. In addition, many display nonlinear properties including thresholds, cascades and surprises (for an authoritative overview, see Steffen et al. 2004).

Is this overwhelming complexity in the Earth system at all possible to steer through institutional design, organizational innovation or novel modes of governance? Or should we instead focus on polycentric approaches as suggested by Elinor Ostrom? How do these approaches evolve over time? And how are these related to the behavior of a dynamic Earth system? These questions have gained prominence the last decade as Earth system scholars continuously elaborate the link between Earth system functions such as the global carbon and nutrient cycle, and human well-being.

In this chapter, I try to elaborate this issue further with a special emphasis on polycentric coordination and 'planetary boundaries'. While the political debate about 'planetary boundaries' and Earth system 'tipping points' is intense (Chapters 1 and 2), my ambition here is to move beyond these discussions, and elaborate how we can come to grips with Earth system complexity from an analytical perspective. I use 'planetary boundaries' as an illustrative case in point, but the argument is intended to be more general and applicable for Earth system interactions in general.

First, I explore the underlying and often misunderstood complexity of 'planetary boundaries'. Understanding the dynamic nature of these 'boundaries', their interplay and associated scientific uncertainties, are critical if we are to understand the institutional challenges posed by Earth system complexity. I then briefly discuss why a simple and conventional institutional analysis is likely to fail in grappling with the institutional dimensions of these proposed thresholds and 'boundaries'. Third, I propose that we need to take a closer look at two governance aspects of 'planetary boundaries' – their underlying and multidimensional institutional architecture, as well as complementary emerging patterns of international collaboration (here denoted as *polycentric coordination*). As I intend to discuss, while polycentric order certainly holds and interesting potential, it is also vulnerable to a range of internal and external stresses in need of further elaboration by Earth system governance scholars.

PLANETARY BOUNDARIES – THRESHOLDS AND INTERACTIONS

If you've ever seen illustrations of the proposed 'safe operating space for humanity' as defined by 'planetary boundaries', it is easy to get the impression that we are dealing with nine independent, quantifiable and static limits. In fact, this is a misperception that has found its way into vigorous public debates about the usefulness of the framework (for example, Lewis 2012, Blomqvist et al. 2012). What is usually lost in the debate is the complicated message that the biophysical thresholds defined by Rockström and colleagues (2009a, b) not only play out at multiple levels (ranging from local–regional to global), but also interact in dynamic ways. For example, as climate change contributes to ocean acidification, these stresses are over time likely to induce the transgression of thresholds in coral reef ecosystems, with detrimental impacts on marine biodiversity. Transgressing the nitrogen–phosphorus boundary can also erode the resilience of marine ecosystems, potentially reducing their capacity to absorb carbon dioxide and thus affecting the climate 'boundary'. And this is just the biophysical part of the issue. Social connections created by trade flows, institutional linkages and technologies such as market information (see Chapter 6), do not make the picture of interactions any simpler.

For example, while there might be few *biophysical* linkages between changes in marine biodiversity and the land boundary, the fact that decreasing fish stocks force fishermen to extract more resources from wildlife and tropical forests (Brashares et al. 2004) is one illustration of how social connectivity also contributes to link 'planetary boundaries' together. Another example: deforestation has for decades been on the international agenda. While forest ecosystems on different continents

might have few biophysical connections, successful conservation policies in one country, has through economic globalization been shown to
lead to increased deforestation in other areas (denoted as a 'displacement
effect' see Lambin and Meyfroidt 2011). Again, governance of 'planetary
boundaries' also implies grappling with connected human–environmental
phenomena, which may embed thresholds at multiple scales. Figure 3.1 is
a visualization of how these interactions potentially play out.

One important conclusion is that the 'boundaries' presently perceived
to be 'safe', could move over time as the Earth system or our understanding of it evolves. Changes in societal values also play an important role
here. Shifts towards more precautionary, or the opposite more risk prone,
values among decision-makers would change the proposed position of
the 'boundary'. This dynamic interplay between systems behavior, values
and politics, should not surprise anyone with insights into the behavior of
complex systems and their decision-making dimensions (for example, Duit
and Galaz 2008).

The multifaceted details of these insights, however, get systematically
lost in the debate. The cognitive challenges in trying to apprehend nested
change at multiple scales, combined with the technical language that
dominates the field are likely to blame here. Nine boundaries are easy to
comprehend – but their underlying multiple possibly nonlinear interactions across biophysical systems, and temporal and spatial scales, are not.
The question for governance scholars is therefore not only whether governance (global, multi-level or polycentric depending on your preferred
theoretical approach) can effectively oversee nine individual 'boundaries',
but also their complex interactions. This is also exactly where the limitations of conventional institutional analysis become clear. The next section
explores why.

AN INSTITUTIONAL ANALYSIS OF 'PLANETARY BOUNDARIES'

The simplest way to analyse the ability of institutions to oversee 'planetary
boundaries' is of course to identify how a wide set of multilateral agreements fit to each boundary. For example, the United Nations Framework
Convention on Climate Change (UNFCCC) is the central international
agreement intended to cover the climate boundary; the Convention on
Biological Diversity (CBD) covers the biodiversity 'boundary'; stratospheric ozone depletion is covered by the Montreal Protocol; and land
use change is partly covered by the Convention on Biological Diversity
(CBD). In addition, some of the boundaries lack an international environ-

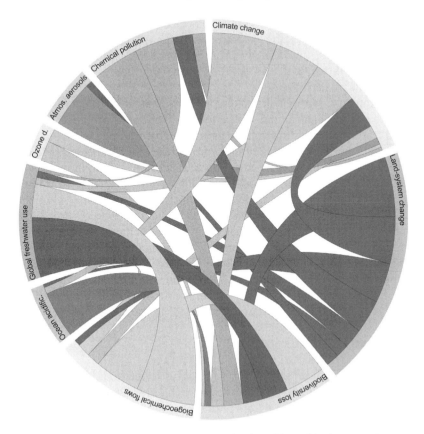

Notes: Impacts of Earth system processes on each other with combined two-way impacts shown in a chord diagram. Values are *first estimations* based on results from a workshop, and range from 0 to 100. Each process is represented by a color. The extents of the processes on the circle are relative to their impact on other processes. Impact factors are only represented relatively, which means that on each process part of the circle the outgoing impact is displayed. Therefore the ending size of a chord on a process displays the impact on the process on the other end of the chord. This helps to identify the processes with the most outgoing impact and gives an easier understanding of the relative importance of certain Earth system processes. It is visible that climate change and land use take up a large share, as they have large impact on other processes. Biodiversity is impacted by many other processes, but has itself less outgoing impact and therefore appears smaller (visualization and explanation by Johannes Friedrich 2013).

Figure 3.1   Planetary boundaries interactions

mental agreement, but are partly covered or at least referred to in various regional agreements – such as biogeochemical flows of nitrogen and phosphorous, ocean acidification and freshwater use. One way to summarize the state of governance and 'planetary boundaries' based on such an institutional analysis is to argue that basically all boundaries already today are captured by the existing environmental institutional architecture. In my mind, this would be a misleading simplification. The reason is quite straightforward, and parallels the saying that 'a chain is only as strong as its weakest link'.

The human–environmental interconnection between 'planetary boundaries' is so prominent that a failure to effectively deal with a few of them could lead to propagating failures in others. For example, while biodiversity and climate change already are part of international processes, a failure to tackle land use change and ocean acidification will undermine them both and trigger nonlinear biophysical change despite partial institutional success in the climate or biodiversity policy arena. A similar argument could be made for the freshwater boundary: the international community might successfully attempt to deal with trans-boundary freshwater issues through the United Nations Economic Commission for Europe's Water Convention, but will systematically fail unless climate change and land use change are addressed effectively. Bluntly put: silo approaches to institutional analysis will undoubtedly fall short in understanding our capacity to govern an interconnected and complex Earth system.

Institutional Diagnostics and Non-Regimes

Institutions have been at the heart of political science analysis of environmental issues in the last few decades. As scholars have explored in great detail, humanly devised rules affect environmental outcomes considerably as they determine the level of cooperation, and the possibility of effective sanctioning of actors which otherwise are likely to over-extract resources (Ostrom 1990, Young et al. 1998, Young 2010).

Creating a new overarching and binding international institutional infrastructure with the ability to link 'planetary boundaries' might consequentially be perceived as an effective strategy. Unfortunately, decades of research on international environmental institutions show clearly that this will be a daunting task.

Oran Young's analysis of the elements that determine whether effective international environmental institutions will evolve at all (Young 2008) can turn any rosy optimist into a discouraged doubter. In short, the problem properties implied by 'planetary boundaries' and their interplay incorporate many of the characteristics that make the emergence of effective institutions extremely difficult. For example, 'planetary boundaries' and their interplay are currently not well understood scientifically, nor are parties such as nation states in agreement about 'their basic character' (Young 2008, p. 122) as the exclusion of 'planetary boundaries' from the meager outcome of Rio+20 illustrates. In addition, efforts to create a new institutional architecture for 'planetary boundaries' is more than likely to impact on 'preexisting institutional arrangements' (Young 2008, p. 123), a fact that complicates the emergence of new international institutions.

One way to denote this lack of institutions to address the interplay between 'planetary boundaries' is 'nonregimes' (Dimitrov et al. 2007). This concept intends to capture a research agenda that analyses the absence of binding international agreements to address urgent global problems. More precisely, these entail 'transnational policy arenas characterized by the absence of multilateral agreements for policy coordination among states' (Dimitrov et al. 2007, p. 231) resulting from nation states' failure to reach a binding agreement due to conflicting preferences, or disagreements on how to define or bound the problem at hand.[22] As an example, while there is an international scientific understanding that coral reefs face severe threats to their resilience and hence ability to provide ecosystem services such as fish protein to millions of people on the planet, nation states have consistently failed to agree to create any coherent and binding international treaty or permanent international organization with the mandate and resources to harness underlying drivers of degradation, or support much needed adaptation measures.

One major obstacle according to Dimitrov et al. is related to scientific uncertainty (Dimitrov et al. 2007, p. 246). Briefly put, coral reef scientists have been unable to provide a full picture of the coral reef crisis' *trans-boundary* character, a knowledge gap that effectively undermines the willingness of sovereign states to create an international institutional infrastructure. Hence despite the possibly critical repercussions a transgression of 'planetary boundaries' and their interactions could have on human well-being, we cannot take for granted that binding international institutions will evolve (Walker et al. 2009).

The role of potential biophysical or ecological thresholds explored in previous chapters is worth restating here. While the risks involved with transgression of a threshold might trigger collective action, any perceived uncertainty about the precise location has been proposed to lead to unavoidable failure to coordinate (Barrett and Dannenberg 2012). In addition, novel problems are known to induce considerable time delays in social responses as decision-makers struggle to make sense and respond to unfolding events – the problem needs to be identified and deliberated with social interests often holding different opinions of the severity of the problem; fragmented and sometimes conflicting information and scientific data needs to be gathered and assessed; and policy options with different distributional impacts need to be assessed and negotiated. Needless to say, 'blame games', path dependent decision-making and political gridlock fueled by scientific uncertainties, can't be discounted (Galaz et al. 2010b, Boin et al. 2005, Scheffer et al. 2003, Harremoës et al. 2001).

So as we enter the Anthropocene, humanity is facing a difficult govern-ance conundrum. Earth system complexity requires novel institutional solutions, and continuous adaptive coordination between sovereign nation states. At the same time the same complex dynamics seriously dilutes the incentives for collective action.

Implications

What does the absence of binding international treaties really imply? And is the term 'nonregimes' really useful for such complex institutional set-tings likely to be found for 'planetary boundaries' and their interplay? The risks involved with the absence of institutions should not be underes-timated. Institutions play not only a key role in governing environmental problems in general, but are also fundamental in coping with the dynamic behavior of any system. Nonlinear environmental change poses a number of generic early warning and response challenges for policy-makers (Galaz et al. 2010b). In short: lack of effective institutions often implies several things. First, that information remains fragmented which leads to 'late warnings' as policy-makers struggle to make sense of scattered warn-ings of a pending crisis ('Is this really going to turn into a big problem'?). Second, their absence also makes it hard for policy-makers to make sense of unfolding events ('What is happening? What is likely to happen next?'), as well as to assign responsibility ('Who is responsible'?). And lastly, a lack of crisis management institutions undermines the ability for rapid coordi-nation, and eventually lays the ground for destructive 'blame games' ('It was not my fault, it was yours/theirs'). 'Blame games' of this sort create well-known obstacles for constructive learning processes, and effective readjustments of existing institutions and policies.

The international 'food crisis' in 2008, which saw soaring food and energy prices, is a good case in point. National and international policy were firstly caught off guard, due to an ineffective capacity to perceive what could have provided tangible 'early warnings' – such as rising oil prices, greater demand for biofuel, multiple severe droughts in wheat pro-ducing 'giants' such as Australia, and a rapid increase in export quotas and trade restrictions. As policy-makers were trying to make sense of unfolding events such as food riots in Haiti, Cairo and Senegal,[23] they also struggled to make sense of the underlying drivers: how much could be related to increasing oil prices and demand of biofuels, financial specu-lation in commodity markets, droughts in Russia and Australia, and lack of investment in agriculture globally? And whose responsibility is it to act in both the longer and shorter term? (see Clapp and Cohen 2009 for details).

To sum up: creating multilateral institutions able to address the inter-connected properties of the Earth system is notoriously difficult. And a failure to effectively oversee these could come with high costs as shocks with poorly understood drivers, propagate across the planet. The question is: if institutional analysis of individual 'planetary boundaries' only takes us so far, what would some useful next steps be?

Next Steps for Governance Analysis?

Global environmental governance entails much more than multilat-eral environmental agreements such as the UNFCCC, or the Montreal Protocol. A focus on governance also implies a broader look at how political actors of various sorts – ranging from nation states, scientific communities, to non-governmental organizations – interact within a wider and multi-level set of formal and informal rules. In addition, the issue is not only collaboration and collective action, but also adaptive govern-ance, that is, actors' collective abilities to adjust and innovate in the face of changing circumstances and 'surprise' (cf. Folke et al. 2005).

In the following sections of this chapter, I discuss two broad areas of research, which I believe need more attention if we really are to come to grips with Earth system complexity, and explore the features of adaptive governance at the international level. First, there is a need to explore the underlying *institutional architecture* and its ability to 'fit' the behavior of a complex Earth system. Second, there is a need to explore the role, potential and limitations of *polycentric order and international actor col-laboration processes*, which potentially could supplement the existing institutional architecture.

EXPLORING THE INSTITUTIONAL ARCHITECTURE OF 'PLANETARY BOUNDARIES'

Political scientists know that institutions matter. The question is how to start an analysis of the complex institutional architecture (defined as 'the interlocking web of widely shared principles, institutions and practices that shape decisions at all levels in a given area of earth system govern-ance' by Biermann et al. 2009, p. 31) taking Earth system complexity into consideration.

Below I discuss four issues worth further attention by governance scholars in the global change community. These are all 'institutional' issues in the sense that they entail the role of formal rules and norms in shaping social outcomes. My emphasis here is on the role of *international scientific assessments, overarching principles, institutional interactions,* and *international organizations.*[24] This is certainly not an all-embracing list, but addresses key functions related to knowledge generation, institutional architectures, and the mandate and behavior of key international actors. In addition, it provides a more extensive governance analysis than those assumed earlier by, for example, Biermann et al. (2012), Young (2010), and Walker et al. (2009).

International Scientific Assessments

Scientific assessments play an important role in international policy-making and institution building. They provide the knowledge foundation on top of which international negotiations can be held; they help define the issues that are on the political agenda at international as well as national and local levels; and they trigger the emergence of international scientific collaboration around certain issues. This is of course only a few of the roles international scientific assessments – such as the Millennium Ecosystem Assessment, the reports from the Intergovernmental Panel on Climate Change, the Global Environmental Outlook, and the World Water Assessment Programme – play in global environmental governance (Mitchell et al. 2006).

The academic community has repeatedly questioned the true influence of international scientific assessments on international environmental policy, especially subfields of international relations where the strategic use of scientific information is seen as only one of many means in the pursuit of power on the international arena (Mitchell et al. 2006, pp. 9–10).

One additional observation is that scientific assessments through their role as platforms for dialogue on global issues also need to consider their internal processes of collaboration, as well as transparency and communication to the outside world, to be listened to as a legitimate voice in the international debate. In the words of Miller (2006, p. 325), international scientific assessments are also 'subject to their own democratic critique'. The setup of the Intergovernmental Panel on Climate Change (IPCC) and the Global Biodiversity Assessment are two of the most prominent examples of international knowledge generation institutions that intend to not only generate policy-relevant knowledge, but also aim to be perceived as transparent, accountable and legitimate by sovereign states and the general public (Biermann 2006).

Issues of transparency and legitimacy of international knowledge generation are likely to become increasingly critical in the Anthropocene. The reason is simple: as insights from Earth system science increasingly enter global political arenas (say, Rio+20), they not only bring to light new global scale phenomena such as interconnected 'planetary boundaries', but also meet conflicting political interests with differing risk perceptions, existing North–South tensions, and debates spurred by unavoidable scientific uncertainties (explored in previous chapters). This is not a trivial observation. Scientific progress in Earth system science is the result of research projects and self-organized scientific meetings and workshops. While these provide an adequate setting from a scientific point of view, they might also lead to a legitimacy gap if the organizational structures and processes are not perceived to respect divergent values and beliefs amongst a broader set of stakeholders (Cash et al. 2003).

The scientific community is of course aware of this gap. Several attempts are currently being made to address this issue through the reorganization of existing global change research programs, into one overarching international research initiative denoted as 'Future Earth' (Mooney et al. 2013). In addition, a number of arenas for cross-system scientific synthesis have indeed emerged the last decade. The Millennium Ecosystem Assessment (2005) provided an important and global collaborative scientific process that holds great potential due to its cross-disciplinary approach and combination of global outlook and regional depth. The Intergovernmental Science–Policy Platform on Biodiversity and Ecosystem Services (IPBES) currently being developed could potentially play a fundamental role in the science–policy landscape for biodiversity and ecosystem services (Larigauderie and Monney 2010). Additional examples include the suggestion to create a science advisory board, or a chief scientific advisor to the UN secretary-general to be led by the UN Educational, Scientific and Cultural Organization (UNESCO) (Nature, *UNESCO to set up UN science advisory board*, 25 June 2012).

Critical issues remain though. How should the institutions that define scientific advisory mechanisms be designed to be viewed as legitimate by external state and non-state actors? How will the scientific community make sure that these bodies capture the nonlinear and often surprising nature of global change despite existing political tensions and conflicting views on the risks of global nonlinear change?

Overarching Principles and Institutional Interactions

Despite the scientific usefulness of defining a 'safe operating space for humanity' by transparent and legitimate knowledge institutions, any discussion about possible institutional solutions at the international level has to acknowledge that these always are the result of negotiations among sovereign states. While a 'new' institutional framework based on the notion of 'planetary boundaries' theoretically could bring some coherence into a highly fragmented institutional landscape, the multilateral development of such a framework is likely to be very slow, or end up being watered down due to the sheer complexity of the issue (see previous discussion on 'institutional diagnostics' and 'non-regimes').

One interesting issue often brought up by scholars from the legal field is to explore the role of *overarching principles in international law* as a way to tackle these difficult challenges. Overarching principles are crucial as they allow international actors such as nation states and international organizations to steer the interaction between different international institutions, and the regulation of norm-conflicts between these institutions. One clear example of an overarching principle is the principle of common but differentiated responsibility (CDR). This principle states that certain risks affect and are affected by every nation on Earth. Responsibilities are differentiated in the sense that not all countries should contribute equally, but instead depending on their wealth.

This principle has had important impacts on international environmental agreements. The 1987 Montreal Protocol for example not only regulated the protection of the ozone layer, but also gave less-developed countries access to financial resources, as well as an extended deadline for coming into compliance. The UN Framework Convention on Climate Change also explicitly embeds the principle of CDR, and the Kyoto Protocol's requirements differed between developed and developing countries (both examples from Stone 2004 who also elaborates some existing tensions in detail). As noted by Frank Biermann (2012, p. 7), the way overarching principles of this sort regulate world trade, as well as concepts of peremptory norms in international law (*ius cogens*, in other words, norms that no state may derogate from) could theoretically provide two good starting points. Similar overarching agreements and norms could theoretically also be conceived for the governance of planetary boundary interactions by guiding the interpretation of international law. Walker and colleagues (2009, p. 1346) for example, argue that '[a]s threats to sustainability increase, norms for behavior toward the global environment are also likely to become part of the *ius cogens* set'. Kim and Bosselmann (2013) argue in a similar vein that there is a legal case for 'a goal-oriented, purposive system of multilateral environmental agreements' based on a new legally binding international norm – a *grundnorm*. The practical implications of these proposals in terms of acceptance and enforcement by sovereign states are highly debatable of course (see Biermann 2012 for details).

The function of these principles in guiding global governance of planetary boundaries is strongly linked to the issue of *institutional interactions* (Gehring and Oberthür 2009). This field tries to analyse the way in which international institutions (for example, the UNFCCC and the Convention on Biological Diversity) interact with other institutions in a more complex setting. As several scholars have pointed out, these interactions are critical for understanding whether individual environmental institutions lead to expected results. For example, early studies in this field showed that the obligations under the World Trade Organization had a 'chilling' effect on negotiations for environmental agreements because they limit the effectiveness of environmental trade restrictions (Gehring and Oberthür 2008, pp. 189–190).

One interesting development in the field is the argument that these interactions in principle can be managed strategically by international organizations to promote environmental policy integration at the international level (Oberthür 2009, Biermann 2009). This is no small opportunity as it would allow for more integrated steering in an institutional setting characterized by serious fragmentation, lack of effective steering of interactions between 'planetary boundaries', and a poor capacity to respond to global change surprises (such as the 2008 'food crisis').

Some suggested strategies that could help manage institutional interactions strategically include the endorsement of inter-institutional learning through joint management among international bureaucracies; expert assessments aiming to promote inter-institutional learning and diffusion; and giving environmental objectives 'principled priority' in cases where environmental and non-environmental institutions are in conflict (Oberthür 2009, Biermann 2009).

A number of critical research issues remain. The practical consequences of trying to create new overarching principles in international law are obviously debatable, and indeed did fail at Rio+20 (Chapter 1 in this book). From a theoretical point of view, no research has looked into the challenges posed to interaction management by the Anthropocene. More precisely, is 'interaction management' possible in a setting characterized by scientifically and politically contested global nonlinear interactions, or are these simply too contested to navigate politically? For which 'planetary boundaries' interactions is this sort of governance more likely, and why?

International Organizations

International organizations (IOs) – such as the United Nations Environment Programme (UNEP) and the United Nations Development Programme (UNDP) – play a key role in global environmental governance as coordinators, knowledge brokers, bridging organizations, and by setting international agendas (Biermann and Bauer 2005, Biermann and Siebenhüner 2009).

Although IOs have been studied extensively the last few years, the emphasis has largely been on their ability to deal with incremental environmental change, rather than nonlinear processes and planetary boundary interactions. The difference between trying to govern individual incremental environmental changes, versus rapid interacting change, is fundamental.

For example, while some implications of climate change and ocean acidification on marine ecosystems can be projected with some certainty, others are likely to unfold as nonlinear social–ecological surprises at multiple levels – such as regional collapses of coral reef ecosystems, and rapid irreversible loss of fish stocks with severe food security implications. This poses a difficult coordination challenge for IOs. On the one hand, dealing with incremental changes in PB (say, coordinating policies to deal with the food security impacts of ocean acidification) require coordinated action evolving around repeated interactions, predictability, and execution by nations, regional organizations and IOs. At the same time, dealing with ecological surprise and cascading effects of environmental change (for example, surprisingly rapid negative shifts in coral reef and agro-ecosystems), require multi-level and ad hoc responses, where a high degree of flexibility and experimentation becomes essential (Folke et al. 2005, Duit and Galaz 2008, Galaz et al. 2010b). Intriguingly enough, the two capacities denoted as *exploitation* and *exploration* seem to be difficult to maintain within the same institutional architecture. The reason is mainly about resources: all organizations need to make tough investment choices between building a capacity to exploit known functions with predictable benefits, or invest in more risky exploration with potentially beneficial but also costly outcomes (Duit and Galaz 2008).

The role of IOs as coordinators and key actors in globally spanning polycentric initiatives hence remains a critical research issue. While the field has made substantial progress in identifying the strengths and weaknesses of IOs in Earth system governance (Biermann and Siebenhüner 2009), much remains to be done in the context of 'planetary boundaries'. How IOs can facilitate the integration of scientific assessments; transform insights from these into actionable information; help formulate policy targets which consider the interplay between thresholds; and facilitate institutional interactions in ways that reduce pressures on 'planetary boundaries' and their contested interactions, are all issues that require scientific and policy attention.

POLYCENTRIC COORDINATION AND INTERNATIONAL ACTOR COLLABORATION PROCESSES

'Governance' implies more than institutions. It is also the result of how actors across a number of levels of social organization are able to build coalitions, share information, extract funding, launch projects, share lessons and act as hubs of innovation. Examples here include a range of global public–private partnerships for sustainable development, internationally spanning networks of cities and municipalities attempting to tackle climate change, and emerging and partly connected regional markets for carbon emissions trading.

These forms of multi-actor and multi-level responses are often viewed as providing *polycentric order* in the sense that they include the self-organizing relationship between many centers of decision-making that are formally independent of each other (this is also the definition of *polycentricity*, see Ostrom 2010).

Polycentricity has gained increased interest from institutional scholars in the field of environmental governance for several reasons. One is that they highlight mechanisms for self-organization that are multi-level and multi-sector in scope (Ostrom, V. 2000), even in settings where formal institutions seem to fail. These mechanisms have hence become interesting in a more general and lively debate about what is perceived as a stagnant UN-process and failing multilateral negotiation processes.

In addition, polycentric order has been proposed to facilitate experimental efforts and learning at multiple levels (Ostrom 2010), a prerequisite for dealing with social–ecological problems that cut across administrative domains (such as the 'problem of fit', Galaz et al. 2008), uncertainty and complex system behavior (Folke et al. 2005, Pahl-Wostl et al. 2007). From a theoretical perspective, polycentric systems could also be viewed as being more robust to external stresses and shocks. The argument is that the institutional diversity in polycentric systems allow the system to recover more quickly after partial failure or collapse, in the same way that biodiversity has been proposed to help ecosystems to bounce back and renew themselves after stresses and shocks (Low et al. 2003).

Despite an increased interest in polycentric order as a potential strategy to deal with complex global environmental problems, we know very little about their features and outcomes (Aligica and Tarko 2011). What is 'polycentric order'? How do these systems evolve over time? In what ways are these forms of governance more promising than formal institutional approaches to environmental governance? And how are these related to the behavior of a dynamic Earth system? These uncertainties make it difficult to empirically assess whether polycentric systems and order at all are an effective and viable strategy to deal with complex global environmental stresses.

In this section, I explore these issues by taking a closer look at how international actors try to deal with the interaction between three 'planetary boundaries': ocean acidification, climate change and marine biodiversity. As I intend to show, while polycentric order certainly holds potential, it is also vulnerable to a range of internal and external stresses.

Processes of Polycentric Order

What characterizes polycentric order?[25] While the question might sound simple, there is actually little agreement in the literature (Ostrom 2010, Ostrom, V. 2000, McGinnis 2000, Aligica and Tarko 2011). A simple way to approach this issue is to focus on four proposed generic processes that underpin polycentric order: information sharing, coordination of activities, problem solving, and internal conflict resolution (displayed in Figure 3.1). As we elaborate in (Galaz et al. 2011b), these processes are important to learn as they can help us understand how polycentric order evolves over time, their structural features as seen from a network perspective, as well as bring to light the 'glue' that keeps this order together.

Information sharing and mutual adjustment is probably the weakest form of polycentric order at the international level. It is 'weak' in the sense that all it requires from actors are investments in creating a joint platform of communication, or routines for information sharing. This process is radically facilitated by recent decreases in costs for information collection and dissemination (Galaz 2009, Galaz 2011). Information sharing is a key component in all types of transnational networks (Andonova et al. 2009) but also supports polycentric order by allowing actors such as international organizations to adjust their behavior to each other in multi-level settings. Ostrom (1999, p. 528) notes that polycentricity allows actors to continuously share information on what has worked in one setting, and therefore support trial and error learning processes. From a structural perspective, a network of relatively loose relations among actors characterizes this 'weak' form of polycentricity. The density of relations among actors is high as the cost of information sharing is relatively low (Yamagishi and Cook 1993). This loose network is focused on information sharing and as such does not, in principle, require any coordinating actor ((a) in Figure 3.2).

However, this form of weak polycentric order is likely to evolve into *informal arrangements of collaboration*, an approach that requires a higher degree of investment and trust than mere information sharing

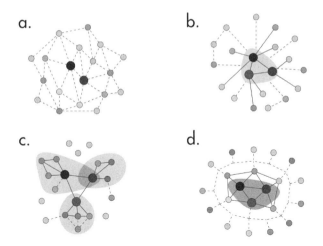

Figure 3.2   Mechanisms of polycentric coordination

(Yamagishi and Cook 1993). These sorts of arrangements are 'informal' as decision-making is coordinated through continuous communication in social networks at multiple organizational levels (Ostrom et al. 1961, p. 841). They remain 'informal' in the sense that they do not result in any formalization of collaboration between international organizations – such as partnerships – but can still be viewed as providing polycentric order through the negotiation of common interest, and repeated coordinated action. The informal nature of this arrangement makes them difficult to pin down empirically, but examples include a suite of 'shadow networks', or 'shadow spaces' that seem to evolve between actors trying to influence prominent institutions (Olsson et al. 2006, Gunderson 1999, Loorbach 2010, (b) in Figure 3.2).

A stronger version of polycentric order ((c) in Figure 3.2) requires a larger investment in *formal partnerships, and coordination of joint projects and experiments*. This can involve a suite of cooperative projects, ranging from investments in monitoring systems, and knowledge production activities, to the deployment of field projects. Joint investments of this kind also open up space for learning and experimentation at multiple levels (Ostrom 2010). The formalization of a partnership requires that social relations change in character from purely information sharing, to relations that require a mutual commitment among actors. Some purely information sharing ties remain, but those actors who do not engage in more costly trust building are eventually left out of the partnership of joint collaboration ((b) in Figure 3.2).

Internal problem solving and conflict resolution ((d) in Figure 3.2) are likely to be the strongest, and most demanding processes of polycentric order (McGinnis 2005, p. 14, Ostrom et al. 1961, p. 838). The reason is that novel challenges (for example, ecological surprise or shocks) forces actors to invest considerable time and resources in interpreting often-conflictive information, and create a shared understanding internally, of what is likely to be dynamic problems with complex and poorly understood drivers (Walker et al. 2009, Galaz et al. 2010b). Hence problem solving and conflict resolution requires enduring ties, built on strong trust, as turbulent or changing conditions create high uncertainty. Here actors need to be able to come together and discuss conflicting ideas and information, and make sense of the changing environment (for example, Olsson et al. 2006). Large groups can provide for a diversity of information and perspectives, but sense making and problem solving can be negatively affected by large group size due to increasing coordination costs (Guimerá et al. 2005, Provan and Kenis 2007).

Polycentric Order in Global Marine Governance

Are these theoretical properties of any use in studying real-world governance challenges posed by the Anthropocene? The answer is 'yes', and the world's oceans are an illuminating example of why. Climate change is already having an impact on marine ecosystems as illustrated by increasing ocean temperatures, and melting sea ice in the Arctic. Oceans are the largest active carbon sinks on the planet and approximately one third of anthropogenic emissions are 'captured' by the oceans. However, this process is changing ocean chemistry, and as the oceans take up more carbon dioxide, they are slowly turning more acidic. A continued decrease in pH could have enormously negative consequences for a wide range of species (Hoegh-Guldberg et al. 2007), and contribute to the bleaching of corals with severe implications for marine biodiversity, and thereby also for the tens of millions of people who depend on living marine resources for their survival (Bellwood et al. 2004, Millennium Ecosystem Assessment 2005).

The global political arena has, despite an ambitious international legal framework embodied in the UN Convention on the Law of the Seas (UNCLOS), and additional international agreements such as the Convention of Biological Diversity (CBD), been largely characterized by a lack of integration among the different policy arenas of marine biodiversity, fisheries, climate change, and ocean acidification (Galaz et al. 2011b, Fidelman et al. 2012). An additional challenge stems from the vast diversity of international players acting at the marine biodiversity–climate–ocean acidification interface (Table 1, from Galaz et al. 2011b). This complex institutional and multi-actor diversity creates a clear need for polycentric coordination and order.

Fortunately, ranges of international actors are trying to address the problems posed by institutional fragmentation and actor complexity. One central initiative is the *Global Partnership on Climate, Fisheries and Aquaculture* (herein referred to as PaCFA), and has been continuously evolving since 2008. It currently includes representatives from FAO, UNEP, WorldFish, the World Bank ProFish Programme and 13 additional international organizations, and can be viewed as an attempt for polycentric order as it involves deliberate attempts at mutual adjustments and self-organized action. In addition, this initiative attempts to address three interacting 'planetary boundaries': climate change, ocean acidification, and loss of marine biodiversity. The question is: how does this sort of global coordination evolve, what are its strengths and benefits, and in what ways do they increase our understanding of more generic governance challenges in the Anthropocene?

A Closer Look at its Evolution

The general feeling of duplication of efforts in marine governance, and the challenges ahead posed by climate change and additional environmental stresses, led some of the international organizations centrally involved in fisheries (notably the FAO, WorldFish and the World Bank ProFish Programme) to embark on a process of identifying gaps and coordination possibilities. As explained by one respondent from a major international biodiversity non-governmental organization (NGO), when asked to explain his view of the policy arena on marine governance and climate change issues:

> If anything, there's too many [organizations]. I mean marine work is very fractured in that you have different industries that are engaged like the oil gas

industry shipping, fisheries, tourism and so on. So it's a whole flora of different sections and they never tend to get together because they have, you know, very different types of specialties and interests and so on. Beyond that there is also the scientists get together in lots of things. We are often quite active in the different scientific fora that pull together different meetings. And in many of these fora we arrange side events, or we host seminars, or we present new technical reports. So it's quite a broad range of things. (Interview from Galaz et al. 2011b)

In March of 2008, in conjunction with a high-level conference on climate change and food security convened by the FAO, an expert workshop on climate change implications for fisheries and aquaculture was organized. The workshop evolved as the result of repeated discussions between a small group of key individuals within the FAO, the World Bank, WorldFish, and UNEP. A wider range of organizations working on fisheries and marine issues were invited to the workshop, including members from academia.

The emerging partnership thus started out as a network of organizations linked through individuals within each organization. In many cases these individuals had worked together in different networks and fora in the past, and thus the formation of a loose partnership based on informal connection lay close at hand. The early link to academia also provided the network with scientific input (Haas 1992). This ambition did not require more than a loose coalition based on continuous communication, and stated interests from organizations ((a), Figure 3.2).

During the process of identifying and pulling together the scientific knowledge on impact pathways and implications and adaptation and mitigation opportunities, the idea of a more formal and powerful partnership arose ((b) in Figure 3.2). Two goals emerged early. The first was to influence the agenda of the UNFCCC COP-15 negotiations. Specifically, PaCFA wanted to highlight the implications of climate change on fisheries and aquaculture, to ensure that adaptation of coastal communities was not left out of the process of allocating adaptation funds, and to get marine issues into the negotiation text. The second goal was to be more active in the field, and to find ways to work together in the partnership to evaluate risks and plan for adaptation strategies of coastal countries. In this initial stage most of the communication was carried out over email, but several meetings, some smaller and some bigger, were carried out with different assemblages of PaCFA members. The partnership members who attended the Bonn climate change talks in June 2009, met in conjunction with this conference to develop a strategy to influence the negotiation text, by joining forces with other groups and linking up with Indonesia. The clear goal and sense of urgency (in other words, marine 'tipping points'), appears to have facilitated the collaboration and communication in the network, and the evolution of communication networks to more formalized patterns of collaboration ((c), Figure 3.2).

The activities of the partnership evolved considerably over the next year, linking informally to a suite of state (for example, Indonesia) and non-state actors (for example, the Nature Conservancy). During 2008–2010, the leadership invested considerable resources in building strategic partnerships, and trying to influence the outcome of the UNFCCC negotiations that took place in Copenhagen in December 2009 (COP-15). COP-15 nonetheless failed to agree on a binding multilateral agreement, and resulted in the Copenhagen Accord. This document does not include any mentioning of marine systems, nor does it direct any of the adaptation funds to projects related to fisheries, or coral reef ecosystems.

The intensity of collaborations and the amount of effort put into the process leading up to COP-15 was high, and the tangible outcome almost non-existent. After COP-15 the network therefore seems to have suffered some fatigue and even disillusionment, which lead members to focus more of their attention on the tasks and agendas of their respective organizations. In addition, as a result of the economic slump at the global level, the likelihood of acquiring funding to expand activities in a way that created more synergistic polycentric order ((c, d), Figure 3.2), was not perceived as attainable in the immediate future.

Hence what at the time before COP-15 was a tightly coordinated network ((c), Figure 3.2) with a joint immediate goal, seems to once again have settled around informal collaboration and information sharing ((a, b) Figure 3.2).

Polycentric Order – Opportunities and Limitations

What can we learn from this particular case study? While studies of international organizations and partnerships are common, my ambition here is to highlight how international organizations of various types attempt to make mutual adjustments and self-organize activities at the local, national and international level. The motivation is not just 'sustainability' in general, but rather to face global and interacting environmental stresses such as those posed by 'planetary boundaries'. Hence the case also highlights some critical and possibly generic opportunities and challenges for polycentric governance of planetary boundaries and their interactions.

For example, some opportunities stand out. Much of the type of collaboration that evolve around the actors builds on loose linkages between individuals, their respective organizations, and a formalized partnership. Loose does not automatically imply weak. Institutional fragmentation in marine governance is significant, and most agencies and organizations are understaffed with limited resources. Many of the members emphasize that PaCFA serves as an important platform for learning and exchange of knowledge, ideas, and information (Galaz et al. 2011b), that is, one fundamental aspect of polycentric order. This also includes what are generally viewed as large international organizations like the FAO and UNEP. Other organizations such as the BCC, but also the UNEP, find that the partnership greatly increases their access to scientific and technical advice, which can facilitate their work and speed up the process of getting important new findings into a policy process.

There is also an interesting multi-level aspect, which is particularly interesting considering that changes in complex systems play out over multiple scales. While collaboration and knowledge sharing seems to be mainly centered amongst organizations at the international level, there are also links to actors working regionally and locally. For example, the WorldFish Centre coordinates a range of projects on the ground, ranging from Asia, Africa, and the South Pacific, while FAO's projects cover practically all regions of the world, implying that learning between international actors could diffuse downwards to more locally placed actors. From a polycentric point of view, these sorts of multi-level linkages could, in a longer time perspective, allow for not only information diffusion and learning across scales (Pahl-Wolst et al. 2007), but also for coordinated action in multi-level governance settings (Brondizio et al. 2009). In line with Haas (1992), global networks like PaCFA can help to support and enable regional and national program development and local initiatives that address planetary boundary interactions. Examples here include the Coral Triangle Initiative (including Indonesia, Malaysia, the Philippines and others), and Partnerships in Environmental Management for the Seas of East Asia (PEMSEA). However, the short existence of PaCFA does not allow for a robust estimate of whether it has been able to truly coordinate local level learning and experimentation.

Despite their tangible potential, the case also brings several serious challenges to the fore. Building a capacity to promote experimentation and learning in a polycentric setting – denoted as 'c' in Figure 3.2 – requires a different type of structure than those loose ties maintained by pure information sharing and ad hoc collaboration. As the partnership has tried to explore that option, several critical challenges emerge. These include the challenge of *keeping the network together*, *negative institutional interactions*, and *lack of resources*.

As elaborated earlier, the frequency of member interactions increases in the time leading up to COP15, and the network moves from a looser format, to a system where connections become stronger, at least among the coordinating group, and there is an explicit focus on several joint outputs. As the goal to influence the UNFCCC process emerges, the ambitions of PaCFA also become more explicitly political. This shift in focus and re-organization (from (a) to (c) in Figure 3.2) started to create some tension within the initiative. Some of the more science-based organizations, for example, felt they did not have the mandate to engage in the 'political game' which PaCFA has now launched itself into. Thus, the inclusion of scientific partners in the partnership – which can be seen as a strength in that it allows for science based capacity building, and the possibility for connecting global and local scales – also presents a challenge as this creates a tension between those that champion scientific legitimacy, and those who want to influence global policies. This tension is far from uncommon for governance networks (Provan and Kenis 2007), and makes it difficult for the initiative to fully evolve into a strong polycentric system ((d) in Figure 3.2).

The sort of polycentric coordination assumed by PaCFA emerges partly to overcome institutional fragmentation and lack of clear steering mechanisms to deal with cross-system interactions. However, this does not make them robust to existing organizational and institutional tensions (Chambers 2008, Gehring and Oberthür 2008). As an example, FAO is one of the primary driving agencies behind the partnership, especially by convening the first critical meetings in 2008 and 2009. While the idea of creating such a partnership had been discussed amongst key individuals at the FAO and WorldFish for some time, the opportunity first presented itself in 2008 when the Committee on Fisheries (COFI), a subsidiary body of the FAO Council, explicitly requested that FAO take steps to 'identify the key issues on climate change and fisheries, initiate a discussion on how the fishing industry can adapt to climate change, and for FAO to take a lead in informing fishers and policy-makers about likely consequences of climate change for fisheries' (FAO 2009, p. 1).

The same member states that gave FAO its mandate to coordinate a global partnership around climate change and fisheries, are nonetheless the same that fail to include marine issues in the negotiation texts related to the UNFCCC. For PaCFA, the failure to achieve some tangible outputs from the UNFCCC COP-15 process seems to have led to lost momentum, and the risk of the partnership dissolving rather than evolving into providing a stronger form of polycentric order. While these sorts of inconsistencies are very common at the international level (Gehring and Oberthür 2009, Oberthür 2009) it is interesting to note how polycentric initiatives which hold potential for dealing with cross-system challenges in a multi-level setting are highly vulnerable to multilateral negotiation processes outside of their control.

Finally, besides a lack of funding for promoting joint projects, the interviews indicate that many of the smaller organizations do not have the funding to attend meetings or set aside human resources to work on joint issues. In the longer term, this is a consideration not only of the impact on the ground, but also of external legitimacy (Provan and Kenis 2007), as financially weaker partners might have problems participating in key meetings and conferences. This seriously hampers the ability of international actors to establish more effective polycentric order by organizing, funding and maintaining collaboration and knowledge sharing mechanisms over time.

One critical question remains: could these sorts of polycentric initiatives really add up in such ways to have a global impact? Or, in other words, can they really, as an aggregate, lead to similar outcomes as would be experienced under strong, self-enforcing multilateral agreements? Answering this question would require additional case studies reviewed over a longer time period, and applying considerable data analysis. Interestingly enough, aspects of this question can be modeled. Vasconcelos and colleagues (2013) present model simulations (in other words, evolutionary game theory) that contrast a scenario where agents try to establish an agreement by which all must abide, and one where smaller groups are established with the aim to overcome shorter-term goals. An important factor here is the agents' perceptions of risk of collective disaster, in other words, in a similar sense as explored for thresholds in previous chapters. The paper is rich in technical details, but the key result stands out: decentralized, bottom-up approaches involving multiple institutions instead of a single global one 'provides better conditions both for cooperation to thrive and for ensuring the maintenance of such institutions' (p. 4). Polycentric arrangements of the type explored here hence clearly can make a difference. Or in the words – 'self-organization may provide a helping hand' (Tavoni 2013, p. 782).

IN SUMMARY

Earth system interactions and their nonlinear repercussions are indeed difficult to tackle. Their unavoidable uncertainty, interplay across levels and across human–environmental systems, and political electricity make them difficult to match to single institutions. And as scholars have shown, effective international environmental institutions do not emerge easily in these settings. This could create serious problems as we move into the Anthropocene dominated by rapid change, nonlinear properties and poorly understood human–environmental change propagating through biophysical teleconnections, and social connectivity such as trade and information flows. This begs the question: How do we sensibly approach this overwhelming complexity?

In this chapter, I've explored the challenges posed by 'planetary boundaries' and their interplay, and suggested that two main analytical paths are possible for those interested in these complicated and novel governance challenges. The first entails taking a closer look at the institutional architecture under which 'planetary boundaries' are currently being overseen. This means exploring the role of scientific assessments, overarching principles, institutional interactions, and international organizations and their joint capacities to address the dynamics of a changing planet. While these could be viewed as tangible policy leverages, their perceptive simplicity should not obscure their respective political difficulties. Scientific assessments have been shown to have a dubious influence on international policy-making; institutional reforms at the international level can be painfully path-dependent; international principles and norms evolve slowly and not without political tensions; and international organizations can only achieve so much with limited resources, skills and mandates.

The second path implies taking a critical look at the emergence of polycentric order at the international level. These evolving global network patterns are embedded in existing formal and informal rules in important ways, but are also the driving force behind global collective action, learning, experimentation, the diffusion of innovation, and adaptability. As the case study indicates, mechanisms of polycentric order ranging from information sharing to coordinated action, do seem to operate at the international level through the interplay between key individuals, international organizations and their attempts to overcome severe institutional fragmentation and actor complexity. An important part of this work is centered on not only attempts to coordinate activities on the field, but also to influence international negotiation processes as a way to secure funding and international acknowledgement.

While polycentric order of this type certainly holds potential (Folke et al. 2005, Ostrom 2010), it is also vulnerable to a number of tensions. For example, even though multi-actor coordination and experimentation would clearly make attempts to create polycentric order more effective, it also creates tensions between actors with different mandates and logics of operating. The case study also indicates that even the weakest form of polycentric order at the international level is dependent on anchoring with more formal negotiation processes. This makes potentially effective collaboration and information sharing patterns vulnerable to unreliable external flows of funding, and negative institutional interactions.

Essentially, these pose severe challenges for the effectiveness of polycentric order at the international level.

Of course, these two paths are not isolated routes. On the contrary, institutional architectures and networked polycentric political action interplay in many ways. The key here is to not lose sight of what these are intended to govern: a dynamic and complex Earth system. We are a long way off understanding what this really implies for the study of governance.

One final observation: the stresses and governance responses explored here have mainly focused on incremental and interacting global environmental change. But complex systems also embed 'surprises', often with hard to grasp cascading effects that rapidly cut across societal sectors and institutional structures at multiple scales. Does this matter for how we analyse the features and impact of governance, and societies' ability to bridge the 'Anthropocene Gap'? The next chapter explores this issue in detail.

4. Epidemics and supernetworks

In the beginning of 2010, I took a short flight from cold Toronto, to an even colder Ottawa to meet with two of the key individuals behind the Global Public Health Intelligence Network (GPHIN): Michael Blench, main technical advisor, and Abla Mawudeku, the network's main Chief. Two things surprised me during our conversation. The first was the incredibly spartan furnishing of the network's headquarters, placed within a huge grey concrete complex hosting the Public Health Agency of Canada. The second surprise was their complete honesty as they described the vast challenges facing GPHIN in the next few years.

I will get back to this interview shortly, but readers should take note of one thing. GPHIN is not just any health network. GPHIN is – without exaggeration – *the* central nervous system of global early warnings and responses to surprising infectious disease outbreaks of international concern. More precisely, GPHIN is at the very epicenter of information gathering, processing and diffusion of early warnings of lethal infectious diseases such as the known highly pathogenic avian influenza H5N1,[26] its less known relative H7N9, the 'new flu' H1N1 (better known as 'swine flu'), and Ebola hemorrhagic fever. Just to mention a few.

Focusing on emerging and re-emerging diseases, and some of their critical governance challenges, might seem like an odd choice for a case study in a book about technology and politics in the Anthropocene. As I intend to show, the capacity of actors at multiple scales to prevent, perceive and respond to surprising outbreaks of novel diseases, provide an excellent illustration of the complex interplay between environmental change, technology and governance.[27]

THE ECOLOGY OF INFECTIOUS DISEASES

In the years 2007 and 2009, Brazil experienced a number of recurring dengue epidemic outbreaks. The symptoms created by the dengue virus are not a very pleasant experience, to put it mildly. Extreme headaches, muscle and joint pain are common and have given dengue fever its second name: breakbone fever. In the worst case, the infection turns out to be dengue hemorrhagic fever (DHF), which results in bleeding appearing as tiny spots of blood on the skin, or larger patches of blood under the skin. In the first four months of 2008 in the city of Rio de Janeiro, there were more than 155,000 cases of dengue fever, more than 9000 hospitalizations, more than 1000 cases of dengue hemorrhagic fever, and 110 deaths, of which nearly half were children (Teixeira et al. 2009).

The puzzling aspect of these outbreaks is that the disease and its vector the *Aedes aegypti* mosquito, were effectively controlled in the 1960s and early 1970s across the whole continent. This was achieved by creating vector control organizations assigned with the task of eliminating breeding sites for the mosquito – such as trash, old tires, clogged rain gutters, containers, street gutters and similar – often by using chemical insecticides (Gubler 1998). In the last few decades the vector and associated outbreaks of dengue fever have re-emerged, and the disease has grown to become the 'second most important tropical disease (after malaria) with approximately 50 to 100 million cases of dengue fever and 500,000 cases of DHF each year' (Gubler 1998, p. 446).

What explains this sudden re-emergence of such a dreadful disease? As explored for many of the cases in this book, causal links for the re-emergence of dengue are highly complex and play out from local to global levels. Several factors seem to be at play: insecticide resistance, extreme weather events, deforestation, land use change, the increased use of non-biodegradable products, and increased social connectivity through trade, travel, and transit. Local circumstances such as insufficient waste management practices, and eroded health infrastructures due to cutbacks in the health sector, are ideal conditions for mosquito breeding and the rapid spread of the disease (Spiegel et al. 2005).

But dengue fever is not the only example of a re-emerging disease. These sorts of intermingled, complex social–ecological drivers are common for many emerging and re-emerging infectious diseases transmitted from animals to humans (known as *zoonoses*) such as influenza A/H1N1, avian influenza (H5N1), Ebola hemorrhagic fever, Nipah virus, and Severe Acute Respiratory Syndrome (SARS) (Patz et al. 2004, Jones et al. 2013).

For some of these diseases, the importance of ecological factors is well established and direct, such as for malaria and its links to urbanization, deforestation and changes in agricultural practices (Gubler 1998, p. 448). For other zoonoses, ecological change plays a smaller but still important role, such as in the case of avian influenza.

Allow me to elaborate: agricultural modification, and losses of wetland due to damming, drainage, and uses for development (such as golf courses, agriculture, and domestic water use), is known to affect the habitat for wild waterfowl. As these habitats shrink, wild waterfowl find themselves forced to share habitation with domestic birds, humans and sometimes even pigs. These changes are problematic from a health perspective as they drastically increase the risks of animal diseases being transmitted within flocks, but also across species and to humans ('spill over'). What's more, viruses mutate due to a process known as 'reassortment' – the mixing of genetic material into new combinations (see Vandegrift and Solokow 2010). Sixty percent of emerging infectious diseases affecting humans today originates from animals (in other words, are zoonotic) (Jones et al. 2008), which implies a connection to ecological systems and change.

Hence emerging and re-emerging infectious diseases are not pure medical challenges that can be solved through improved disease tracing, quarantine, and mass-vaccination. Infectious diseases of this sort are also inherently ecological challenges, resulting from complex social–ecological interactions, which require broader and more diverse governance approaches. Before I explore that issue in-depth, we need to understand one fundamental governance challenge closely related to the main issues explored in this book: the need for early warnings and prompt polycentric coordination in the face of surprise, connectivity, and the possibility of threshold change.

EARLY WARNING, EARLY RESPONSE

The necessity for early and reliable warnings of pending epidemic out-breaks has been a major concern for the global community since the creation of the very first international health regulation in the mid-nineteenth century, when cholera epidemics overran Europe. The 1851 International Sanitary Conventions marked the beginning of international governance on infectious disease. One major problem has been how to tackle the fact that there are strong disincentives for individual states to report disease outbreaks to the international community. The reason is quite straight-forward: any information about disease outbreaks affects the domestic tourism industry, threatens exports, and results in rapidly increasing political pressure and 'blame-games' that undermine public trust of governments (Morse 2007, Institute of Medicine and National Research Council 2008).

The implications for the global community of chronic under-reporting of novel infectious diseases are obviously not trivial. Just consider the potentially devastating human costs of a pandemic with high rates of mortality (Rubin 2011). This is particularly critical in the case of this book, as coordinated responses are needed in order to avoid the transgression of critical epidemic thresholds. Once these thresholds have been transgressed, control and containment measures become essentially meaningless due to the rapid spread of the disease.[28]

As explored in Chapter 3 about 'planetary boundaries', and later in the cases about algorithmic trade and geoengineering, coordinated responses are considerably more complicated than generally assumed. The reason is that both early warning and responses typically engage a wide range of public and private actors at multiple levels of social organization. At the global level, these actors include the World Health Organization (henceforth WHO), the World Organization for Animal Health (OIE), the Food and Agriculture Organization (FAO), Health Canada and its GPHIN system, and the United States Centers for Disease Control and Prevention (CDC). At the national level, epidemics engage not only affected national and local governments, scientific communities, including veterinary medicine and epidemiology, but also non-governmental organizations, such as Doctors without Borders (Médecins Sans Frontières, MSF). In many cases, WHO plays a critical facilitating and coordinating role. The organization's capacity to coordinate global responses to urgent epidemic threats was nevertheless seriously questioned and challenged by other international organizations such as the World Bank, throughout the end of the twentieth century (Fidler 2004). Its ability to bounce back from this critique, and transform its way of operating in the face of epidemic surprise is an excellent illustration of the interplay between technological change, governance and complexity.

Recovering from Surat

The failure of the WHO to respond to a devastating outbreak of bubonic plague in 1994 in the rapidly urbanized Indian mega-city Surat, is often mentioned as the hardest blow towards the credibility of the organization as a coordinator of global health emergencies. The local newspaper *The Hindu Universe*'s vivid report from the first unfolding days of the epidemic is a vibrant reminder of the chaotic nature of disease outbreaks:

> [P]eople fleeing the affected zones are heading in all directions and taking the hysteria with them. With the discovery of three people afflicted with plague in a Bombay hospital, panic has gripped that city as well. Tetracycline, an antibiotic for plague treatment, has disappeared from chemist shops not only in Bombay but also in Delhi.[29]

As the crisis escalated, WHO did not take action other than issuing press releases, and organizing one official visit to Surat more than two weeks (!) after the first media reports. Although less than 60 persons died from the plague, the outbreak led to economic losses estimated to US$260 million in Surat alone, boycotts and travel restrictions against India, as well as highly speculative allegations in national media of the plague being the result of covert US genetic engineering experiments, whose results were intentionally released by a rebel group in Kashmir (Lin 1995, Garrett 1996, Dutt 2006). The critique against the WHO was devastating.

In 2003 – less than 10 years later – WHO was able to meet the early warning and response challenges posed by the previously unknown infectious disease 'SARS' (severe acute respiratory syndrome). When SARS – a then unknown respiratory infectious disease which has led to the death of 813 persons – slowly but surely spread across the world, from southern China to Hong Kong, Vietnam, Canada, Singapore, and the United States, WHO was able to swiftly disseminate information and coordinate laboratory networks, national health agencies, and non-governmental organizations to contain the spread of the disease (Heymann 2006, van Baalen and van Fenema 2009). A similar story can be told for the surprising outbreak of the 'new flu' A/H1N1 (more often known as 'swine flu', Brownstein et al. 2009).

The difference in the speed of response between the 1994 bubonic plague in Surat, and newer outbreaks of international concern is clear, but still somewhat surprising (for a more quantitative assessment, see Chan et al. 2010). What contributed to the transformation of WHO, from a heavily criticized, and clearly ineffective international organization, to the main coordinator of global responses to infectious disease outbreaks?

THE INFORMATION REVOLUTION IN GLOBAL EPIDEMIC GOVERNANCE

An important part of this transformation seems to be linked to the rapid evolution of information and communication technologies. Put bluntly, advances in information processing capacities have allowed states and public health experts to sidestep the then outdated WHO information dissemination model that built on voluntary reporting from members states to the organization.

This evolution started in the mid-1990s with the creation of ProMED, a moderated, email-based reporting system. ProMED started as a simple mailing-list among a small group of people who were interested in emerging infectious diseases. As Larry Madoff – the current editor and director of ProMED explains it:

It attracted some publicity which led to more and more people joining and it grew virally from about 40 people sharing emails, to hundreds then thousands

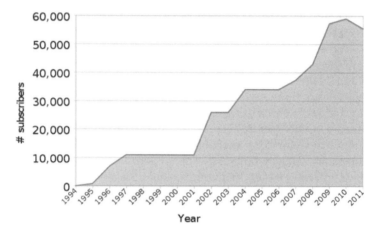

Notes: Based on Madoff and Woodall (2005), and personal communication with ProMED Editor Larry Madoff. The decrease in 2011 indicates a removal made by ProMED staff of non-active email accounts.

Figure 4.1   Number of ProMED subscribers

over several years. ProMED was the brainchild of three very devoted individuals. Since it did not have a very strong organizational or financial basis, it was not assured of continuity. In 1999, it was brought into the International Society for Infectious Diseases, which provided it stability and a more robust structure.

ProMED has grown considerably over time (Figure 4.1), and allows health experts all over the world to rapidly share information about disease events via a moderated email list, thereby effectively bypassing official and considerably slower channels of epidemic information dissemination (Mykhalovskiy and Weir 2006, Morse 2007). (Interestingly enough, an email list among divers helped marine scientists uncover the spread of coral bleaching in the 1990s, see Galaz et al. 2010a.)

A second fundamental information technological innovation was the Global Public Health Intelligence Network (GPHIN), today located in Ottawa. This early detection system was developed for the WHO by Health Canada in the mid-1990s. GPHIN is based on 'web crawlers' – software programs that automatically and methodically browse the World Wide Web in search for particular key words, such as 'unknown pneumonia', 'respiratory disease'. GPHIN gathers information about unusual disease events by monitoring Internet-based global media sources such as newswires, Internet websites, local online newspapers, and public health email information services in nine languages. GPHIN is able to detect the first hints of about 40 percent of the 200–250 outbreaks subsequently investigated and verified by the WHO each year (Mykhalovskiy and Weir 2006). As I will explore later in this chapter, the filtering process is the result of an intriguing combination of computer and human intelligence.

THE EMERGENCE OF SUPERNETWORKS

Improved information processing is only part of the picture. As I have explored in Galaz (2009) and Galaz (2011), the rapidly decreasing costs of information gathering, analysis and distribution, seems to have triggered the emergence of very large-scale, diverse (in terms of membership, resources and competence) and connected networks. Bluntly put, there seems to be a rapid increase in connectivity between the very diverse set of international and national actors involved in epidemic early warning and response. I denote these emerging patterns of collaboration in global epidemic alert and response 'supernetworks' – that is, connected 'networks of networks', consisting of international organizations, state agencies, universities and non-state actors such as NGOs.

The term 'supernetwork' might seem superfluous in a field replete with theoretical perspectives such as 'network society', 'policy networks', 'social networks', 'actor-network theory', 'network management', and 'global knowledge networks' (for example, Castells 2009, Ansell 2006, Klijn and Koppenjan 2004). The attractive feature of the term 'supernetwork' is its emphasis on the steering challenges posed by complexity and uncertainty; the recognition that networks are interconnected at multiple levels; and the notion that information technology plays a key role in understanding the dynamics of these networks (Nagurney et al. 2006). More precisely, the emergence of 'supernetworks' is the result of multiple and diverse investments in disease monitoring systems, information sharing collaborations, and coordination mechanisms for health interventions and projects.

Note that these sorts of polycentric coordination patterns – of varying degrees, ranging from pure information sharing to strong mechanisms for collaboration and conflict resolution – also were observed in Chapter 3 on governance aspects of 'planetary boundaries'. The networks here, however, are considerably larger in size. International organizations such as the WHO and associated programs such as the Global Influenza Programme, the World Organization for Animal Health (OIE), the FAO, the US Center for Disease Control (CDC), the Red Cross, Médecins Sans Frontières and others, are embedded in an almost overwhelming web of partnerships, programs and collaborations.

The network visualization shown in Figure 4.2 illustrates the linkages between different organizations involved in early warning and response to

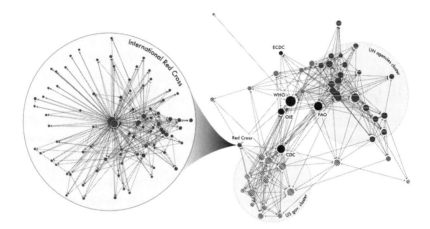

Notes: The nodes in this network map represent official web-pages associated with an
actor (say, WHO.int for the WHO), and links are hyperlinks between different nodes (say, a
link between WHO.org to the US CDC homepage on avian influenza). The analysis builds
on three basic assumptions. The first is that all organizations – national, international,
state, non-state, private, civil and so on – advertise their activities on the Internet. Second,
they selectively choose which other actors to link to. That is, organizational web-pages
do not hyperlink randomly, but rather selectively based on how central and trustful they
consider the actor they are linking to is (see Park 2003, McNally 2005, Bruns 2007 for
details). For example, if both the World Health Organization's and the US Center for
Disease Control's projects on avian influenza list the Red Cross as a partner, but none link
to the World Wildlife Fund, that probably means that the Red Cross is viewed as a more
central partner. For this map, the starting points were key international avian influenza
early warning and response programs.

Figure 4.2   Avian influenza supernetwork

avian influenza at the global level. The method applied here is what has
been denoted as 'hyperlink analysis', a network mapping strategy that uses
web pages, and the way these are linked to each other (and how often),
as a very rough proxy of existing collaboration patterns (see Park 2003,
McNally 2005 for details of methodology).

One interesting observation from the map is the relatively strong
representation of US agencies (bottom-left). For example, even though
the WHO and the OIE are identified as central 'bridging' partners, the
network illustrates the existence of an important UN–US collaboration. This pattern is expected: it is well known that the US has been a
strong advocate of enhancing the global preparedness against EIDs, and
has invested considerably in different monitoring systems (Institute of
Medicine 2008, Bravata and McDonald 2004).

A critical reader might rightfully ask: are these not merely global partnerships displayed as global networks? My answer would be that emerging patterns of global governance of emerging and re-emerging infectious disease, are better understood as 'supernetworks', rather than just international collaboration or global partnerships. Allow me to elaborate: even though an NGO such as the International Red Cross might seem to be marginal at the global scale compared to others actors such as the US CDC and the WHO (Figure 4.2), collaboration within the Red Cross on emerging infectious diseases is in *itself* a vast global network of actors. As displayed in the same figure, zooming in to the Red Cross component of the 'supernetwork' opens up another set of linkages and nodes, but this time with quite a diverse participation from additional international actors. As I explore next, the ability to process information through these supernetworks, solve specific problems through 'collective intelligence', and coordinate their action in times of crisis, illustrate that these are more than simple global partnerships.

THE EXPANSION OF SUPERNETWORKS

A suite of very diverse epidemic surveillance, laboratory and response networks do seem to have emerged in the last few decades with a different focus, geographical scope, and membership (Galaz 2009). Hyperlink maps of this sort are a quick way to grasp and visualize some of these complex actor networks. These do not by themselves prove an *actual* collaboration pattern between actors, in the same way that linkages in blogs does not prove tangible collaboration between the authors.

This still does not capture the interlinkage in 'networks of networks'. A closer look based on interviews with key actors in international epidemic early warning and response (see also Galaz 2009, Galaz 2011) reveals that these epidemic surveillance and laboratory networks not only expand in parallel, but also overlap, and embed intriguing patterns of information sharing and coordination. This is no coincidence, but seems to be linked to advances in information and communication technologies. For example, David Heymann, the main architect of WHO's current early warning and response structure to emerging and re-emerging diseases, summarizes the impacts of GPHIN on their operations in the following way in an interview that I held with him in Geneva in March 2008:

> All of the sudden, we had a very powerful system that brought in much more information from more countries, and we were able to go to countries confidentially and validate what was going on, and if they needed help, we provided help. And we provided help by bringing together many different institutions from around the world that started to work with us. In Kikwit [Kongo-Kinshasa],

we had about 20 different NGOs working with on the Ebola outbreak. And we invited more and more NGOs to come and work with us, government agencies, and what not. And we ended up with an outbreak alert and response network that was formalized in the year 2000.

Hence as information gathering and processing capacities increase, WHO saw a need – and probably also opportunity after decades of harsh criticism from member states – to expand its collaborations with both national governments and NGOs, and act as the key facilitator of global epidemic responses (see also Fidler 2004). Patrick Drury, coordinator of WHO's Global Outbreak Alert and Response Network (GOARN), elaborates the issue in this way:

> There is a bigger 'networks of networks' that David [Heymann] talks about, these would bring in many of the same partners. In GOARN you have CDC, and UNICEF, MSF and Red Cross. Which you also have in the different coordination groups for meningitis vaccine and yellow fever vaccine. Or in global polio eradication. These are enormous, but some are very small and you would bring in the global influenza with laboratories and national influenza centers. But that is the network of networks which has no substance, no defined substance. It's there, the function, but in a highly chaotic, very undefined way.

The WHO's Global Influenza Surveillance Network, a network created already in 1948, is an excellent example of expanding supernetworks. This network has increased its member base over time, especially after global epidemic crises. As Kenji Fukuda, the coordinator of the WHO's Global Influenza Programme, explains in an interview when I met him in Geneva at the WHO headquarters in 2008:

> It was from 1948, at the very beginning of the network. So the network was just like a centre, the world influenza centre, plus a few laboratories, well this is the start of the network in 1948. Then, the network was expanding itself and now we have 122 National Influenza Centres in 94 countries and it is still expanding. The goal of this type of expansion work is that we hope that the WHO surveillance could cover globally. [. . .]The virus could appear anywhere so the surveillance needs to cover as much as possible and also the network needs to be functioning as efficiently as possible so that, you know the goal of this network could be reached.
> *Question from me*: There have been a few high-profile outbreaks; you have SARS, you have repeated cases of Avian Influenza. Did you see an increased interest in joining the network after these sorts of events?
> Yeah definitely, I think the emergence, well the sort of expansion of H1N1 from 2004 onwards has really sort of catapulted influenza into everyone's attention and that has really driven a renewed interest in both global surveillance for these viruses but also participation of countries so they can have an Influenza Centre in their own countries.

It should be noted that this expanding early warning and response capacity is far from being just global, or coordinated purely by the WHO. On the contrary, there are a range of similar and parallel attempts to expand collaboration between existing monitoring and response systems by NGOs. Dan Sermand at Médecins Sans Frontières – with over 15 years of experience of response operations in the field – says in relation to how activities are coordinated on the ground when internal MSF networks are unable to cope:

> *Question: What if you are facing some uncertainty of the disease? How do you coordinate your networks?*
> Each time we have a suspecting case of fever, or something very wrong, the first thing we do, is that we contact WHO. Immediately. So we give you the sample, say the stool sample, or the blood sample, or whatever, and we work with them immediately. [. . .] So the collaboration is, I will say, immediate. We are not waiting for the people to start to die, to smell that something is smelling shit, and that something could be . . . So there is immediate collaboration, so we call them and 'send you the sample with the first plane', or the first car or whatever. So, 'please go on with your laboratory and tell us what's going on'. That is systematic.
> *Question: So that is not formalized?*
> No, no, but it's not personal. WHO knows that we will always call them if we are suspecting things or something is very bizarre.

This sort of partly formal and informal multi-level collaboration – in the form of information sharing, and coordination of complementary capacities – is not at all uncommon for 'supernetworks' in this domain (Galaz 2011, van Baalen and van Fenema 2009). The incentives for all participating partners are clear: local and national actors gain from collaboration with outside actors as they bring in resources and competence, and international actors such as the WHO and FAO strengthen their role and legitimacy as coordinators of health emergencies. In other terms, the very *raison d'être* of international organizations such as the WHO and FAO stems from their ability to facilitate multi-level polycentric responses. From an information theoretical perspective, this puts very high requirements on international organizations to gather, evaluate and disseminate information. This ability is not centralized, but builds fundamentally on effective collaboration with its partners. Allow me to elaborate this somewhat.

COLLECTIVE INTELLIGENCE IN SUPERNETWORKS

Early detection is much more than scanning for reports on the Internet, or gathering reports from member countries and labs. It also entails the capacity to facilitate collaborative processes of *collective intelligence*. This term was coined in the late 1990s by French philosopher and media theorist Pierre Levy, who wanted to capture the capacity of people to solve specific problems, through self-organized collaboration in large virtual networks (Levy 1999, Heylighen 1999, Jenkins 2006). A simple and iconic example of this type of large-scale and distributed form of problem solving is the *Polymath Project*.

On 27 January 2009, Timothy Gowers decided to put an unresolved mathematical theorem on his blog – the Hales–Jewett theorem. The interesting thing with the post was not the theorem in itself, but rather the aim to prove it collaboratively using blogs and a wiki as the main collaborative tools. Over the next month, '27 people contributed approximately 800 substantive comments, containing 170,000 words'. The progress of this process was surprisingly rapid. On 10 March of the same year, Gowers announced that the collaboration – including the participation of highly respected mathematicians, academics, as well as a high-school teacher – had managed to find the proof (from Gowers and Nielsen 2009).

My argument here is that 'collective intelligence' also seems to be a phenomenon that grows out from expanding, overlapping and loosely coordinated epidemic networks. The collaborative analysis of the then new infectious disease SARS in 2002–2003, is one clear example. The global health community's ability to 'dodge the bullet' (Michelson 2005) built on the ability of the WHO to facilitate not only response, but also epidemic problem solving. As epidemiologists and health officials struggled to understand the new disease, the first laboratory analyses showed that this respiratory disease did not originate from normal nor avian flu. But if it isn't flu, then well, what is it? (see Greenfeld 2006 for an excellent inside look into the workings of the epidemic intelligence teams).

The Global Outbreak Alert and Response Network (GOARN) worked to bring together 13 laboratories in 9 countries, along with more than 50 medical clinicians in 14 countries, to identify the causative basis for the disease, its mode of transmission, and possible responses (Michelson 2005, see also Heymann 2006, van Baalen and van Fenema 2009).

Additional examples of this type of distributed problem solving in large-scale networks include the analysis of and response to an unknown form of influenza in Madagascar in 2002 (Eurosurveillance 2002), investigation of febrile deaths of unknown cause in young adults in Bangladesh in 2001 (Hsu et al. 2004) and prevention of epidemics of yellow fever in Côte d'Ivoire in 2001 and Senegal in 2002 (Heymann 2003). This process of problem solving – supported through tele-conferences, continuous email exchange, and secure web-pages – involves all features elaborated above: the capacity to process information of unfolding events (Poutanen et al.

2003); the ability to support distributed problem solving by maintaining secure communication channels; and the skills to bring together actors within supernetworks to respond in a coordinated fashion (Michelson 2005).

Human–Computer Interactions[30]

Another interesting feature of supernetworks can be found in critical monitoring systems such as GPHIN. The acknowledged successes of GPHIN are not only based on the network's technical capacities alone, but rather on the interplay between data mining technologies, algorithms and human analysis. This interplay is essentially what makes the system able to filter out considerable 'noise' (in other words, rumors, disinformation, unrelated reports), and create tangible and useful early warnings to the international community and members in supernetworks. The way analysts at GPHIN were able to spot the first warnings of the 'new flu' A/H1N1 is a good illustration of this interplay.[31]

Initial local news reports flowing in to the GPHIN system – after a sophisticated machine translation and semi-automated filtering process (see Blench 2008) – indicated a minor outbreak of respiratory illnesses in villages in southern Mexico. Articles in local newspapers suggested that these could be traced to environmental contamination from a pig farm. However, the GPHIN analyst in charge of making risk assessments of incoming reports felt increasingly unconvinced about the newspapers' claim. As an excellent example of serendipity, she was familiar with the region of interest and decided to dig deeper. Going back years in news reports about reported disease outbreaks in the same area, she noticed that this type of respiratory illness was a unique event that required further investigation. After some deliberation with her bosses in the organization, her intuition induced further investigations and eventually, a formal request by the WHO to the Mexican government to investigate the sources of an unknown outbreak of respiratory diseases within its national borders.

While this might seem like an unusual strike of luck, it illustrates one of the key features of GPHIN's success: the ability to combine advances in information processing capacities, with human analysis based on not only epidemic expertise, but also cultural understanding, and local knowledge. For example, analysts are recruited and trained to be able to pull out epidemic early warnings from sources as diverse as local business news reports of increased sales of vinegar in parts of China used as a natural remedy for common cold and flu symptoms; company reports bragging about increased sales of antiviral in certain parts of the world; to Chinese (sic!) financial news of decreased sales of poultry in African countries, due to rumors of avian influenza.

It should be noted that the information from GPHIN not only links to the WHO, but also provides important input to additional agencies and international organizations such as the Food and Agricultural Organization and its multi-organizational monitoring program Global Early Warning and Response System for Major Animal Diseases including zoonoses (GLEWS). Systems like GLEWS integrate reports from GPHIN with those from ProMED, and similar warning systems like Interlink and Argus, and in rare cases, even individual blogs. Again, combining computer-processed incoming reports with human analysis, based on extensive contextual knowledge, is critical for an effective filtering process.

The Limits of Collective Intelligence

It should be noted that this problem solving process is partly collaborative, but also partly competitive. While there certainly is a common interest in trying to isolate and sequence the virus, fierce competition between internationally renowned labs is common. This requires some exquisite negotiation skills from the side of the coordinators.

For example, in the midst of an unfolding epidemic of a novel avian flu virus denoted as H7N9 in April 2013, Chinese researchers uploaded the genetic sequences of viruses isolated from the first three human cases to a research database – the Global Initiative on Sharing All Influenza Data (GISAID). Sharing of influenza data is critical as it allows participants in supernetworks to separately monitor drug resistance, track the evolution of viruses, and develop diagnostics and vaccines. As the Chinese team was about to submit their first academic paper including an analysis of the sequences for the virus to the *New England Journal of Medicine*, they were informed that several other research teams were well underway to do so – based on the Chinese group's uploaded data. What's more, private companies in collaboration with US agencies were already in the process of developing H7N9 vaccines, without informing or inviting the Chinese group to participate (Butler and Cyranoski 2013). This sparked serious tensions between the Chinese researchers and US-based researchers, government bodies and private companies Novartis in Basel, Switzerland, and the J. Craig Venter Institute in Maryland.

This is not the first time international conflicts about information sharing of the genetic sequence of viruses emerge. Indonesia, for example, has for a long time supplied H5N1 virus samples to the WHO Global Influenza Surveillance Network for analysis, and preparation of vaccines for the world by commercial companies. However, the country questioned the model correctly pointing out that the produced vaccines would be unavailable to the country during a pandemic. In 2007, Indonesia decided to suspend its sharing of viruses with the WHO, a decision that created considerable tensions within the international health community (Fidler 2008).

How recurring and successful phenomena of collective intelligence are within supernetworks is difficult to quantify. It should be noted that any attempt to interpret incoming information about pending epidemic crises, does involve some element of distributed problem solving, facilitated by information and communication technologies. More precisely, making the choice of whether to post or discard information about a potential outbreak of concern not only requires advanced algorithms, data-mining technologies, and individual analysis skills, but also the ability to contact and facilitate problem solving between collaborators around the world, with different expertise, resources and interests. These are the processes that allow actors in fragmented settings to navigate around epidemic thresholds, respond to epidemic surprise as rapidly as possible, and mitigate potential cascading impacts.

AN IMPORTANT NOTE ON CONTROVERSIES AND NEGLECTED DISEASES

Prompt early warnings, collective intelligence, and prompt multi-level collaboration between members in loosely connected supernetworks, might sound like a perfect governance model for complexity and surprise. Such a simple analysis would miss two important issues: One is the role of *controversies*, and the second is the need to acknowledge *neglected diseases*.

As Melissa Leach and colleagues at the STEPS Centre in the UK wisely point out, international responses to epidemics are continuously beset with critical controversies (Forster 2012, Dry and Leach 2010). These can be found in all parts of the long chain of crisis response – ranging from early warning, to epidemic analysis, and response.

This brings again us back to Ottawa, and the meeting with Michael Blench and Abla Mawudeku. Managing information about epidemic outbreaks always has arduous diplomatic dimensions – no countries like it when external actors (such as GPHIN) without their consent, distribute epidemic information to the world about events within their territories. Something as simple as naming a disease is controversial. For example, naming the 2009 A/H1N1 influenza 'swine flu' aggravated pork producers as they saw their income nose-dive in the aftermath of China, Russia and Ukraine having banned pork imports from Mexico and parts of the United States.[32] The alternative 'Mexican flu', was seen as stigmatizing by Mexico (understandably),[33] and analysis of virus samples are – as explored above – both collaborative and highly conflictive processes.

In addition, coordination of responses is not always straightforward due to unavoidable uncertainties about unfolding events – how fast is the disease likely to spread? Where? What is the most effective response, and by whom? The intense debates about the WHO's different pandemic 'phases', the pandemic alert and associated national responses such as mass vaccination for the outbreak of A/H1N1 ('swine flu'), is an excellent example of how controversial responses to epidemic surprise can be (Doshi 2011). Epidemic models – such as mathematical models trying to predict the spread of avian influenza – are critical tools for decision-makers as they try to get an overview of possible outcomes when planning for interventions. At the same time, models always embed assumptions and unavoidable uncertainties, making them disposed to conflict with the conclusions from other models. At worst, model assumptions lead to policy-conclusions that seriously clash with local realities. For example, in Gabon in 1995–1996, American and French Ebola control measures were perceived as so inappropriate and offensive by villagers that, when international teams arrived to address a further outbreak there in 2001, they met fierce local armed opposition (Leach and Scoones 2013, p. 14).

One additional critical observation is that not all zoonotic diseases gain the same high level of attention from the international community as avian flu and additional new forms of influenza. On the contrary, some of the worst zoonotic diseases that affect the most vulnerable communities in the world remain inadequately monitored and poorly responded to. In 2012, Delia Grace and colleagues mapped the interlinkages between zoonose hotspots, and poverty. The analysis not only maps out the geographical spread of zoonotic diseases such as Rift Valley fever, HIV–AIDs, influenza, malaria, measles and dengue. It also notes that many of the zoonotic diseases which affect poor communities the worst – such as *brucellosis* and *Trypanosomosis* (sleeping sickness) – also are those diseases which face the most serious monitoring, and response challenges (Grace et al. 2012).

These so-called 'neglected diseases' (Yamey 2002, see also Moran et al. 2009) are widely unreported, or seriously under-reported, which makes timely and effective responses, such as those explored earlier for supernetworks, almost impossible. In the case of *brucellosis* for example – a disease that infects humans as well as cattle, swine, goats, sheep and dogs – under-reporting is flagrant. The disease causes flu-like symptoms in humans, including fever, weakness, malaise and weight loss, but it is not easily transferable between humans. As Grace and colleagues note, for every 1 million cases of animal infections in East Africa, less than *one* case is reported to the World Organization for Animal Health (OIE). The situation is similar for other diseases reported to OIE. In short: 'When there are 999,999 missed cases for every one report, surveillance is not fulfilling its purpose' (Grace et al. 2012, p. 18).

While several examples of successful initiatives to tackle neglected diseases do exist (Molyneux 2004), it is also clear that neglected zoonotic diseases not only have severe negative impacts on the health of poor communities in large parts of the world, but also have received only modest attention compared to more high-profile diseases which are seen to threaten international security, such as avian influenza (Scoones 2010, Elbe 2010). This so-called 'securitization' has managed to pool in large investments (about 1.5 billion US$ according to the World Bank, in Elbe 2010) in preparedness efforts that have driven an important part of expanding supernetworks.

Note that this does not imply that current monitoring systems for pandemic threats are flawless. On the contrary, reports on serious gaps in monitoring (for example, for the avian virus in poultry, and pig influenza) are common (Garten et al. 2009, Butler 2012). And as I write this (end of May 2013), the central decision-making body of the WHO, the World Health Assembly, is discussing reallocations in the organization's budget that would result in more than halving of the funding available for outbreak and crisis response. These two limitations should be kept in mind as I in the next section summarize the interplay between complex systems behavior, institutions, and the role of supernetworks.

SUPERNETWORKS, POLYCENTRIC COORDINATION AND INSTITUTIONS

Emerging infectious diseases (EID) such as animal influenzas (for example, 'avian influenza' and 'swine flu') are illuminating examples of the difficult governance challenges posed by complexity and connectivity in the Anthropocene. As the human enterprise modifies landscapes, expands urban centers, modifies the climate, industrializes livestock, swine and poultry production, and increases global connectivity through trade and transport networks – it also transforms disease risks. Complexity and connectivity are abundant in this issue area: epidemic thresholds, nonlinear changes in ecosystems, possible epidemic surprise as genetic material in influenza virus reasserts, and the prevailing risk of rapid transmission and propagation over national borders. I would like to conclude by highlighting two issues that will be of value as we move to later parts of this book:

the role of macroculture in network collaboration, and the fundamental role played by information technological change.

Macroculture and Narratives

It is interesting to note how global responses have evolved over time. As I have explored in this chapter, global health governance has seen the rapid expansion of 'supernetworks'; loosely coupled and collaborative networks of networks. These networks expand as responses to perceived crises, and seem to embed a whole array of mechanisms for polycentric coordination – quite opposite to previous arguments of the information revolution 'swamp[ing] us in information', where 'we often devote more time to managing information and less to producing new high-quality ideas' (Homer-Dixon 2002, p. 26). These entail sophisticated information processing capacities; implementation of joint cross-organizational projects; coordinated epidemic responses across levels of social organization; and the support of processes of collective intelligence – distributed, virtual problem solving. This sort of robust collaboration might seem surprising considering the ever-prevailing risks of collective action failures, as well as the contested nature of epidemic outbreaks and responses.

The presence of a shared 'macroculture' seems to play a key role here. As Rod Sheaff and colleagues (2010) note, the expansion of health networks can only provide a tangible option for governance if relationships in networks build on trust, reciprocity and mutual interest (cf. analysis in Chapter 3). In addition, the rules of engagement need to be known and clear to all parties. The authors' note that a 'network's "macroculture" articulates these shared goals and values, giving them concrete representation in artefacts, language and symbols' (Sheaff et al. 2010, p. 780). Several forms of artifacts are possible, such as products, services, technologies, and symbols such as logos. They also include explicit rules of engagement and collaboration, as well as implicit non-negotiable values. More precisely for this case, a common sense of urgency and need for response before the transgression of critical epidemic thresholds; a spirit which highlights the joint benefits of international collaboration; and artifacts in the form of technical guidelines, and technological tools such as GPHIN and ProMED, unquestionably play a role in allowing actors to create a shared sense of urgency, and tap into the economic and intellectual resources of supernetworks. Put simply, concerns about the implications of transgressing epidemic thresholds and resulting cascades, drive cooperative action facilitated by macroculture.

But there is another and more critical view on the impacts of macroculture, and its institutional implications. Melissa Leach and colleagues

(2010) note that while the framing of the problem in terms of an 'outbreak narrative' might seem logical, it is only one of many possible delineations of system boundaries, strategies and type of interventions. Narratives of this sort do not exist in a vacuum, but are in general advanced by organized political interests – ranging all the way from international organizations, to government agencies, non-governmental actors and business actors. In sum: 'The narrative therefore frames the system in global terms. It focuses on a particular interpretation of disease dynamics (sudden emergence, speedy, far-reaching, often global spread) and a particular version of response (universalized, generic emergency oriented control, at source, aimed at eradication). More subjective dimensions include the value placed on protecting global populations, which often implies protecting particular populations in richer countries' (Leach et al. 2010, p. 372, see also Dry and Leach 2010).

Supernetworks and International Institutions

It is also interesting to note how these loosely coupled networks interplay with more formal international institutions, such as the International Health Regulations. It is well known that international institutions play a critical role affecting zoonotic risks, for example through trade agreements, or by failing to regulate land use changes. These institutions also shape state and non-state action as they set the rules for coordinated global responses. For example, current requirements on nation states to report to incidents of disease outbreaks of international concern are laid out in a very detailed way in the present International Health Regulations.

The role of supernetworks is slightly different. Perceived epidemic threats (such as SARS and avian influenza) spur the emergence of global and connected networks. Over time, these create a pressure on international institutions to adapt to what they perceive as critical global health challenges. For example, the emergence of novel zoonotic diseases (such as avian influenza) is intrinsically linked to the effectiveness of a suite of institutional rules at multiple levels, for example though urbanization, land use change and technological development. The potential of these diseases to rapidly transgress dangerous epidemic thresholds creates incentives for international cross-organizational action, in this case through the emergence of global early warning and response networks. This collaboration evolves *despite* malfunctioning formal institutions (in other words, the International Health Regulations before the year 2005). As nation states eventually agreed to reform the International Health Regulations in 2005, the revisions built on the technical standards, organizational operation procedures and norms developed by these networks *years in advance*

(Heymann 2006). In theoretical terms, global networks have the ability to promote adaptability in international institutions, and thus can help combat 'institutional arthritis' (Young 2010, p. 382, see also Eddesson 2010).

Technology, Networks and International Politics

As should be clear by now, advances in information and communication technologies play a key role in this problem domain. Advances in monitoring technologies through, for example, semi-automatic data mining of online news (with GPHIN being the most prominent example) has provided the international health community with a set of powerful tools to bypass severe obstacles posed by under-reporting induced by national self-interest. It has also allowed centrally placed actors such as the WHO and FAO to expand their membership base in existing networks, grow in regions where surveillance and laboratory capacities have been weak, and provided the necessary tools to facilitate prompt collaboration and problem solving at multiple levels – all the way from local doctors affiliated to international NGOs combating Ebola hemorrhagic fever in Central Africa, to the very top of the WHO outbreak preparedness and response headquarters.

The importance of controlling the tools that allow harnessing of information should not be underestimated. As noted earlier, the very *raison d'être* of international organizations stems from their ability to contribute to coordinated responses to what is viewed as global challenges. From an information theoretical perspective, this not only implies very high requirements on information gathering, evaluation and dissemination capacities, but also allows a few organizations to become the main coordinators of global epidemic governance. As Castells (2009) noted, information and power are intrinsically related.

It should be noted that the focus in this chapter differs considerably from previous attempts to unpack the role of information technological change in international politics and governance. For example, the focus of Mitchell (1998) was on the role information and transparency plays in supporting the effectiveness of international regimes. Others, such as Castells (2009) and Morozov (2012), explore how new forms of coercion evolve in conjunction with changes in the information processing capacities of companies and states. And lastly, organizational scholars such as Markus and colleagues (2000), and Kurland and Egan (1999) have elaborated the emergence and organizing principles of 'virtual organizations'. My focus here is on how decreasing costs of information processing and dissemination, and large investments in global response capacities, lead to the expansion of global 'supernetworks' with intriguing features of interest for governance scholars.

FINAL THOUGHTS

The 'supernetworks' explored here embed many of the properties which characterize adaptive modes of governance in the face of complexity (Folke et al. 2005). While providing an interesting case of global coordination in polycentric settings, the case also illustrates the contested nature of epidemic governance in the Anthropocene. Policy-makers not only have a hard time addressing underlying drivers of emerging and re-emerging diseases, but also find themselves in controversies about definitions ('swine flu', 'Mexican flu' or 'new flu'?), surveillance disputes, and disagreements about the most effective responses. In addition, the existence of 'neglected diseases' also illustrates the global health community's limited capacity to tackle endemic, but important, zoonotic diseases with severe impacts on human well-being.

Hence governing 'tipping points', connectivity and surprise, are much more than simple management issues. They are intensively political and controversial matters, but which nevertheless require collaboration in highly polycentric settings. This dilemma becomes just as clear in the next case study chapter – geoengineering technologies.

5.   Engineering the planet

Imagine a room with climate change and biodiversity experts discussing the future of planet Earth. Or more precisely, whether it might be justifiable to deploy so-called geoengineering technologies which rapidly could cool the planet, but with considerable and unquantifiable environmental risks. As the discussions unfold over several cups of coffee, and effortlessly move between ethical, technical and political dimensions, one of the experts raises his voice irritably. The very foundations of the discussions are now being questioned. 'Surely the comparison can't be between a world where massive geoengineering technologies are deployed, and an ideal world where humanity suddenly manages to drastically cut emission of greenhouse gases? The fair comparison is between a world moving towards a catastrophic +3°C to +4°C increase in global mean temperature, and one where the global community strategically mitigate some of its worst impacts through geoengineering interventions. If we frame the comparison in that way: which alternative would be most beneficial for biodiversity, ecosystems and humans?' There was a moment of complete silence. And I know this because the discussion did in fact take place in London, at an expert meeting hosted by the United Nations Convention on Biological Diversity in June 2011.[34]

In this chapter, I elaborate the intriguing governance challenges created by the development of geoengineering technologies – another illuminating example of the 'Anthropocene Gap'. Several books, journal articles and essays in glossy magazines have been written about geoengineering the last few years (for example, Hamilton 2012, Goodell 2010, Fleming 2010). The approach here is different: My emphasis is on the intricate governance challenges posed by emerging and converging technologies as we enter a new geological epoch. I explore regulatory gaps and the complex actor constellations in this domain, as well as the poorly understood trade-off between innovation and precaution in a new setting characterized by rapid and nonlinear environmental and technological change. But first, a very brief exploration of the geoengineering debate.[35]

HACKING THE CLIMATE

As historians are quick to point out, proposals to modify Earth's climate are not new at all. On the contrary, history provides us with a number of examples of imaginative weather modification schemes, and larger scale suggestions on how to rework global scale climate dynamics. Proposals of the latter type can be both amusing and fascinating to read, like Roger Angel's paper 'Feasibility of cooling the Earth with a cloud of small spacecrafts near the inner Lagrange point (L1)' (2006) in the highly respected scientific journal *Proceedings of the National Academy of Sciences*. In summary, Angel's article elaborates the possible use of electromagnetic acceleration and ion propulsion to eject very thin meter-sized 'flyers' with such speed, that they manage to escape Earth's gravity. These would then effectively create a 'sunshade' over the planet. As the author bluntly notes: 'They would weigh a gram each, be launched in stacks of 800,000, and remain for a projected lifetime of 50 years within a 100,000-km-long cloud. [. . .]. The concept builds on existing technologies'.

Suggestions to deploy large-scale technological interventions as a means to combat climate change that a decade ago would have been discarded as science fiction are slowly moving toward the center of international climate change discussions. The expert deliberations hosted by the Convention on Biological Diversity mentioned in the opening of this chapter, is only one of many examples of how debates on the usefulness of geoengineering technologies slowly but surely, are sliding into international policy arenas. Another example is the joint inquiry on geoengineering carried out by the Science and Technology committees of the US House of Representatives and the UK House of Commons (House of Commons 2010). And lastly, the initiative by the United Nations Intergovernmental Panel on Climate Change (IPCC) to include an assessment of geoengineering technologies in its fifth Assessment Reports (AR5).

What is Geoengineering?

It is almost like a law of nature: any expert meeting on geoengineering is likely to start with heated discussions on what the concept really means. I choose a simple strategy by simply defining *'geoengineering' as intentional, technological large-scale interventions in the climate system to mitigate the impacts of anthropogenic climate change*. It should be noted that several definitions with different emphasis can be found in the literature (for example, Royal Society 2009, US Government Accountability Office 2010, Bipartisan Policy Center 2011).

Definitional issues aside (and we will return to them shortly), geo-engineering technologies often include proposals which imply techno-logical interventions which either (1) remove carbon dioxide through, for example, ocean iron fertilization, carbon capture and storage, afforesta-tion and reforestation, and the enhancing of soil carbon through pyroly-sis of biomass (in other words, 'biochar'), or (2) enhance our ability to regulate incoming solar radiation by deploying fine solid particles in the stratosphere (so-called 'stratospheric aerosols'), space mirrors, or whiten-ing clouds in the lower atmosphere (for a more elaborate presentation, see Royal Society 2009, GAO 2011).

Several issues make geoengineering an incredibly slippery concept to grapple with and discuss from a governance perspective.

The first is that these technologies in many cases are not mature, ready-to-launch-artifacts, as we normally would perceive the term 'technology'. On the opposite, the term 'geoengineering technologies' includes a set of very different types of proposals and ideas with very different degrees of technological maturity – that is: (a) technologies which only exist or have been tested on smaller scales (such as biochar, or fertilizing oceans with iron); (b) proposals which build on natural observations of natural events and computer simulations (such as observations of reduced global temperatures after volcanic eruptions, and the proposed injection of strat-ospheric aerosols); and (c) pure concept development (for example, space mirrors). The 'geoengineering' portfolio is – simply put – diverse and in continuous development as the result of scientific developments and tech-nological change.

Second, the opportunities and risks involved with the potential large-scale deployment of geoengineering technologies are just as diverse as the contents of the portfolio. Several reports and articles have already elaborated the opportunities and risks associated with the deployment of geoengineering technologies (Bellamy et al. 2012, Royal Society 2009, Williamson et al. 2012). As a simple and general rule, interventions which aim to regulate incoming solar radiation – such as the injection of sulfur particles in the stratosphere – are the only type of interventions which would have an immediate cooling effect on the planet in case of a per-ceived 'climate emergency'. These are also those entailing the highest risk of unwanted consequences as they could lead to not only ozone depletion by creating free radicals which catalyze detrimental chemical reactions, but also induce changes in the behavior of major weather systems such as the South East Asian monsoon.

Geoengineering technologies which instead aim to actively remove carbon dioxide – such as reforestation, enhancement of soil carbon and chemically based carbon dioxide capture and storage ('artificial trees') – are considered safer, but are generally slow-acting meaning that their cooling impacts would take decades to observe. Very bluntly put: rapid impact technologies are risky, slow impact ones are safer.[36]

Third, the word 'intentional' in any definition of geoengineering is important as it excludes several global environmental change phenomena that characterize the Anthropocene (Steffen et al. 2007). For example, even though the global modification of water vapor (Dessler et al. 2008) and particle emissions from power plants (Wild et al. 2007) have clear climate impacts, these phenomena are normally not considered as geoengineering because they lack explicit *intent* to intervene in the climate system. We will return to this important issue later in this chapter.

Fourth, technical definitions and politics are entities which have proven hard to separate. As witnessed during the 2010 meeting of the parties of the Convention on Biological Diversity (CBD) in Nagoya (Japan), the choice on what – and more importantly what not – to include in the geoengineering portfolio is not only a scientific choice, but in many senses also a political one. For example, how large-scale must a reforestation/afforestation project be to be defined as 'geoengineering'? And should carbon capture and storage really be considered 'geoengineering technology' rather than a more advanced form of mitigation? National interests play a clear role in this ongoing definitional battle (Sugiyama and Sugiyama 2010, see also *Earth Negotiations Bulletin* Vol. 9, No. 595, p. 15).[37]

In sum: 'geoengineering' remains a notoriously slippery, scientifically contested and politically charged concept. It also works as an umbrella term for technologies with an immense span of maturity, ranging from essentially computer assisted thought-experiments to small pilot experiment facilities (see for example, discussion in Heyward 2013). Why should we even bother trying to govern something so incredibly amorphous, contested and uncertain? Two recent geoengineering experiments illustrate why it is hard to ignore an in-depth discussion on these issues: the SPICE hose experiment, and 'rough' geoengineering in the form of ocean iron fertilization.

A Giant Hose called SPICE[38]

In 2010 the British research consortium SPICE (Stratospheric Particle Injection for Climate Engineering) was granted £1.6 million to test the feasibility of constructing a 1 km long hosepipe, attach it to an immense helium balloon, and then elevate it. Once that first extremely challenging engineering problem was solved, the team would pump around 150 liters of water droplets into the atmosphere. This field experiment was an integrated part of a larger research project with the ambition to elaborate the engineering challenges and public perceptions associated with stratospheric aerosol geoengineering. That is, in the longer term, the research team foresaw the design of a 20 km long hose, and specially designed particles which in principle could be pumped out through the hose, thereby cooling down large regions, or the whole planet. In the words of Principal investigator, Bristol University's Matt Watson: 'It's a Plan D [. . .]. It's essentially an insurance policy for the situation where we hit a tipping point in climate change quickly'.[39]

Even though the project had gone through multiple elaborate reviews as requested by its public funders, the official public launch of the plans drew immediate negative attention from environmental organizations. After the date and time of the test was announced, more than 50 organizations spearheaded by the Canadian NGO ETC Group and the online campaign 'Hands Off Mother Earth' (H.O.M.E.), signed a petition forcefully objecting to the experiment. Their worry was that the experiment was a 'Trojan Hose' – that is, that it would lead to the development of a technology that not only entails considerable risks if deployed on large scales, but that also would provide politicians with an excuse to avoid tough decisions on reducing greenhouse-gas emissions. In the midst of this debacle, one of the study's main funders decided to postpone the trial for six months to allow the researchers to take stock and reexamine the social dimensions of the project. A few months later the research team announced that the SPICE field experiment had been canceled. The reason was not primarily public opposition, but instead according to the principal investigator, the 'lack of rules governing such geoengineering experiments', and a perceived conflict of interest created by the fact that two researchers associated with the project, had a patent application pending for an 'apparatus for transporting and dispersing solid particles into the Earth's stratosphere' by 'balloon, dirigible or airship' (from Cressey 2012). This implied a clear conflict of interest that had not been communicated to the funding agencies beforehand. NGOs around the world celebrated, but several critical questions lingered: should these sorts of experiments be allowed and funded with public funds? And if not, who is entitled to make such a decision?

Rough Geoengineers at Haida Gwaii[40]

It was not the first time, nor is it likely to be the last. In July 2012, a small corporation called the Haida Salmon Restoration Corp. (HSRC) decided to launch one of the largest and most controversial geoengineering experiments in history. Even though several small-scale ocean fertilization projects had been assumed before – about 13 according to UNESCO

(Wallace et al. 2010) – this particular project was considerably larger. The corporation used ships to dump more than 100 tons of iron sulfate, plus iron oxide and iron dust, into the ocean 320 km off the coast of Haida Gwaii in the northwestern coast of British Columbia. The rough general idea behind ocean iron fertilization geoengineering is that the iron will trigger carbon dioxide consuming plankton blooms. These would later both boost marine food webs, and later sink to the deep oceans leading to a net removal of carbon dioxide. As reported by the *Guardian*, the iron seemed to have generated a plankton bloom as large as 10,000 square kilometers, visible from satellite images.

This asserted salmon restoration project unleashed intense international discussions of the true intentions and legal status of the experiment. Clearly, this was more than an attempt by a Canadian First Nation village to give what one of their representatives had called 'the marine environment a vitamin supplement'. One of the reasons for this suspicion was the restoration project's explicit connection to Russ George, a Californian businessman. George had made himself an international reputation as one of the leading proponents and entrepreneurs of ocean iron fertilization as a geoengineering option through his former 'ocean seeding' company *Planktos*. But more importantly, early news reports stated that Russ George seemed to have convinced the village that the project would not only help restore salmon stocks, but also would pay for itself in the form of carbon credits to be paid out through international carbon markets. Representatives of the Old Massett Village Council representing a community of 700 people, allegedly helped fund the project with $2.5 million through the Gwaii Trust Society, and a village reserve fund. This re-payment through carbon credits will of course, never materialize (Tollefson 2012).

Criticism was immediate and furious as international media spread the news about the experiment a few months later. The timing of the news release is interesting here: the *Guardian*'s news-breaking article on the Haida Gwaii experiment was released in mid-October 2012, in the midst of heated negotiations at the United Nations Convention on Biological Diversity (CBD) conference of the parties in Hyderabad, India. One of the topics on the agenda: the regulation of geoengineering technologies, where governments of Bolivia, the Philippines and several African nations as well as indigenous peoples' organizations, called for stronger regulation of these technologies.

Several non-governmental organizations questioned the legal status of the acclaimed restoration project and maintained that it violated at least two different international agreements – the Convention on Biological Diversity's 'moratorium' on geoengineering experiments, and the London

Convention's regulation on ocean dumping. The United Nations' International Maritime Organization expressed 'grave concern' over the project, and noted that ocean fertilization due to its entrenched ecological risks, should not be allowed except for legitimate scientific research.

As international debate and critique built up, so did pressure on Canadian government agencies to clarify how much they really had known about the experiment in advance. Data rapidly became an important issue in this discussion – had the experiment really led to carbon sequestration? How much in that case? And could the company verify its claims that salmon stocks had started to recover? The company's repeated claims that it had collected massive amounts of very valuable data – or '180-million data points' – spurred not only anger but also curiosity amongst scientists. But in March 2013, officials from Environment Canada headed to the corporation's Vancouver headquarters and other locations, and seized scientific data, journals and files. At the end of May of 2013, the media reported that Russ George had been removed as a director of Haida Salmon Restoration Corp., and that the Old Massett Village had decided to initiate a strategic review of the project. Russ George on the other hand, claimed that the village did not have the authority to remove him from the leadership of the company.

In the tumult, one quote from one of the First Nation representatives, Old Massett Chief Councillor Ken Rea, stands out as a vivid illustration of the sort of conflicting perceptions that emerge from the 'Anthropocene Gap'. In a press conference arranged by representatives from the Haida Salmon Restoration Corp. at Vancouver Aquarium in October 2012, Rea vigorously asserted in defense of the project: 'On a changing planet, we need to take bold steps and the people of Old Massett believe this is the right step'.[41]

GEOENGINEERING GOVERNANCE – DEBATES AND OPTIONS

The SPICE and Haida Gwaii events illustrate a number of things. The first is that the development of geoengineering technologies are much more than nonsensical science fiction fantasies which can be ignored. On the contrary, the issue is indeed very much on the table, and the intense conflicts between private actors, governments, international organizations, and non-governmental organizations needs to be addressed by political means. Second, the cases exemplify the diverse forms of 'technologies' and actors interacting in this emerging political arena. More precisely, SPICE illustrates how scientists and government research programs attempt to explore solar radiation management technologies, whereas Haida Gwaii is an example of how entrepreneurs tap into private funding sources in the development of carbon dioxide removal technologies which mimic 'natural' processes in the ocean. And lastly, the examples bring to light the myriad of regulating national and international actors, and the lack of a clear, overarching governance architecture able to effectively resolve conflicts around funding, patents, experimentation and (if considered needed) large-scale deployment. Allow me to briefly elaborate this last point (for a more detailed discussion, see Burns and Strauss 2013).

Institutional Fragmentation and Regulation Gaps

As for all of the cases explored in this book, institutional fragmentation and actor complexity is considerable in this issue area. As Karen N. Scott, professor in law, elegantly explores the issue (Scott 2013), while there currently exist a few legal instruments which *explicitly* apply to geoengineering, there still exist a myriad of international legal principles which impose obligations on states if an activity is seen to create a 'significant risk of serious harm to the environment' (p. 330). These include – amongst others – the principles on prevention of harm; the obligation to protect vulnerable ecosystems and species; the precautionary principle; and principles of cooperation, information exchange, and environmental impact assessment (Scott 2013, p. 330. The details about their legal applicability for geoengineering activities remain contested).

The lack of a legal framework that explicitly addresses geoengineering technologies has spurred a number of initiatives by international organizations to contribute to a more coherent and detailed institutional architecture. The International Maritime Organization (IMO) and current modifications of the London Convention and Protocol, is one important attempt to regulate ocean-based geoengineering technologies such as ocean iron fertilization (Bodle et al. 2012, p. 124). The Convention on Biological Diversity (CBD) is another critical organization embedded in an institutional framework that has made considerable attempts to regulate and frame international debates about geoengineering. For example, in 2010 the parties of the Convention agreed on a decision inviting states to: 'in the absence of science based, global, transparent and effective control and regulatory mechanisms for geo-engineering [. . .] that no climate-related geo-engineering activities that may affect biodiversity take place, until there is an adequate scientific basis on which to justify such activities' (Decision X/33). It should be noted that this decision, incorrectly referred to as a 'moratorium' on geoengineering, is not as such legally binding which implies a lack of enforcement powers (Scott 2013, pp. 332–333, Bodle et al. 2012, pp. 123–124). At the same time, the fact that the decision represents the consensus of 193 Parties sends an important political signal (Bodle et al. 2012, p. 123), and plays a key role in the heated public debates triggered by SPICE and the Haida Gwaii experiments.

Actors and Networks

Governance is more than legal principles: It is also about the actors and their patterns of collaboration as they try to influence possible institutional futures. The numerous international actors, their differing perceptions, preferences and responsibilities are worth mentioning here. The following is far from a complete list, but let us start with the most prominent international institutions that in different ways engage with the topic. These would include not only the already mentioned organizations IMO and CBD, but also the United Nations Framework Convention on Climate Change (UNFCCC), the Intergovernmental Panel on Climate Change (IPCC), the United Nations Environment Programme (UNEP), United Nations General Assembly, the United Nations Educational, Scientific and Cultural Organization (UNESCO), the World Meteorological Organization (WMO), and the Intergovernmental Oceanographic Commission (ICO) (from Bodle et al. 2012, p. 123ff).

Other key actors at the national level that have been critical for bringing geoengineering to the fore of international politics include the UK House of Commons and the US House of Commons Select Committee on Science and Technology. They have been 'critical' by engaging in a unique joint inquiry on geoengineering technologies which resulted in several synthesis reports, consultations and hearings between selected officials and geoengineering scientists (for example, House of Commons 2010, United States Government Accountability Office 2011).

In addition to these formal political bodies, you will also find vocal and important non-state actors. Amongst the most prominent ones I would like to mention: *private actors* – including numerous actors ranging from smaller carbon-dioxide removal start-ups, ingenious individual inventors, to large philanthropists like Bill Gates and Richard Branson; *think tanks* – such as the Bipartisan Policy Center, and the Copenhagen Consensus Center; *environmental NGOs* – such as the ETC Group, and Friends of the Earth; and *epistemic communities* – such as NOVIM, the British Royal Society, and the Solar Radiation Management Governance Initiative.

The list is by no means complete of course, and the most interesting aspect from the perspective of this book is how all these political players, entrepreneurs, and intellectuals intermingle and create networked alliances within a highly complex institutional landscape. Here is one illuminating example.

Science, Private Interests and Governance

In 2009, the Royal Society, a highly prominent British scientific institution, published what is likely to be one of the most quoted reports on geoengineering technologies: 'Geoengineering the climate – science, governance and uncertainty'. The importance of such an established scientific organization in helping to bring geoengineering as an issue to the public and political eye, is difficult to overstate. For example, as the UK House of Commons explored possible collaboration options with the US House of Commons Committee on Science and Technology Committee, geoengineering rapidly became the gravitating subject towards which both organizations perceived that they could collaborate. The reason was the timely release of the Royal Society geoengineering report and its recommendations on the need for further investigations (House of Commons 2010, p. 44). This UK–US collaboration has as already mentioned, led to numerous reports by these two institutions, again raising the profile of the issue, thereby providing an important opinion forming function. Over time, journalists have asserted that additional and important Royal Society reports, as well as some key experts, have been partly funded by well-known private interests: that is, by Bill Gates and Sir Richard Branson through their philanthropic organizations.[42]

An important part of the story is that both Gates and Branson have economic interests in companies that build on the development of geoengineering technologies such as *Carbon Engineering* (http://carbonengineering.com/), and *Intellectual Ventures* (http://intellectualventureslab.com). Environmental NGOs – such as the ETC Group, and scholars of geoengineering governance – were quick to point out the problems created by this explosive intermix of commercial interests, individual geoengineering scholars, scientific institutions, and ongoing regulation debates where these experts systematically provide advice. As Jane Long, director for the Lawrence Livermore National Laboratory (US) formulated it: 'We will need to protect ourselves from vested interests [and] be sure that choices are not influenced by parties who might make significant amounts of money through a choice to modify climate, especially using proprietary intellectual property'.[43]

Governance Options

Several governance options to close a number of these 'regulatory gaps' (Reynolds 2011, p. 121ff, Bodle et al. 2012) exist, an issue that has been explored by several scholars recently (for example, Victor 2010, Virgoe 2008, Lin 2009, Banerjee 2011, Bodansky 2011).

One critical choice is whether a governance model should build on existing international agreements (for example, CBD, UNFCCC, London Convention), or instead on a new overarching institutional framework. There are also different positions on whether this latter option should be based on voluntary codes of conduct, or on a stronger and binding international agreement (see Banerjee 2011 for details).

Another critical choice is what *function* an existing governance arrangement is expected to have. As explored in different 'governance scenarios' by Harvard scholar Daniel Bodansky (2011), international agreements and mechanisms can be designed to govern very different challenges associated with the development, experimentation and deployment of geoengineering technologies. These include a legal setting intended to fully bring the development of geoengineering technologies to a halt (what Bodansky in quite a revealing way denotes 'premature rejection'); institutional mechanisms which support secure and transparent funding streams from government and private actors ('inadequate funding'); or rules designed to avoid a situation where a rich private actor, or a few countries, goes ahead with the large-scale deployment of geoengineering without global consent ('Greenfinger' and 'Unilateral state action'). Geopolitical dimensions are critical in this last model (see Blackstock 2012).

Whatever option and scenarios the international community passages into, I would like to note the emerging polycentric patterns of collaboration (as explored in Chapter 3) that seem to evolve also in this issue domain. More precisely, international geoengineering scholars build global collaborative networks for information sharing and learning (for example, the *Solar Radiation Management Governance Initiative*); epistemic communities explore and coordinate around their preferred governance options and principles (for example, the so-called *Oxford Principles* see Rayner et al. 2013); international organizations intentionally build in cross-referencing between international agreements to reduce institutional fragmentation (such as those between the CBD and the LP/LC); and non-state actors such as environmental NGOs, private companies, philanthropists and think tanks collaborate within their respective networks in their attempts to shape the future of institutional development.

PLANET ENGINEERS

Very little of what I've said so far is new to those familiar with the geoengineering debate. But the previous sections provide an important point of departure as we dwell deeper into three additional issues, which I believe are critical illustrations of the 'Anthropocene Gap'.

Farewell to Precaution?

The 'precautionary principle' is probably one of the most well-known principles in environmental law. With its acclaimed roots in English common law in the nineteenth century, and the German legal principle *Vorsorgeprinzip*, the principle has been enshrined in numerous international treaties and declarations, and provides the basis for European environmental law (Cameron and Abouchar 1991, Foster et al. 2000, Kriebel et al. 2001). The precise definition and interpretations of the principle differ (VanderZwaag 1999), but have been argued to have a set of common core components: taking preventive action in the face of uncertainty; shifting the burden of proof to the proponents of an activity; exploring a wide range of alternatives to possibly harmful actions; and increasing public participation in decision-making (Kriebel et al. 2001, p. 871).

But the precautionary principle is more than a technical legal principle intended to guide institutional development and environmental policies – it is a principle that also deeply connects to 'the philosophical and spiritual relationship between humankind and the environment which sustain our physical existence' (Cameron and Abouchar 1991, p. 2). This certainly calls for more complex elaborations of the underlying philosophical logic of the principle (Jensen 2002, Gardiner 2006). More interestingly for the purpose of this chapter, is how the precautionary principle is used by scholars, international organizations, environmental activists, think tanks and others, to support different positions and preferences over institutional alternatives, in the geoengineering debate.

The precautionary principle might at first glance seem to inevitably imply a careful approach to the development of geoengineering technologies considering their risks, such as changes in regional precipitation patterns. This interpretation of the principle can be found in the Convention on Biological Diversity, and its decision to limit the use of geoengineering technologies (decision X/33), as well as under parts of the London Protocol which regulate ocean-based geoengineering (article 3.1, see Bodle et al. 2012, p. 48 for details). It is also repeatedly used, for example, by key NGOs such as the ETC Group, Greenpeace and Friends of the Earth, just to mention a few.[44]

While this application of the precautionary principle intuitively makes sense, legal scholars have noted that the complete opposite interpretation is also possible. For example, under the Convention on Biological Diversity, the precautionary approach is introduced in the preamble, where it is noted that 'where there is a threat of significant reduction or loss of biological diversity, lack of full scientific certainty *should not be used as a reason for postponing measures to avoid or minimize such a threat*' (from Bodle et al. 2012, p.48, my italics). A similar formulation of the precautionary principle can be found in the Rio Declaration from 1992: 'Where there are threats of serious or irreversible damage, *lack of full scientific certainty shall not be used as a reason for postponing cost-effective measures to prevent environmental degradation'* (Principle #15, see Dodds et al. 2012 for a review).

As Bodle and colleagues (2012) elegantly elaborate on the issue, this means that the 'precautionary principle' – one of the key pillars of environmental law – actually both opposes and supports the development of geoengineering technologies depending on the interpretation (see also House of Commons 2010, p. 34 for a similar interpretation). In the former case, these technologies should not be developed nor tested in field trials due to their possible negative unexpected environmental consequences. In the latter case, lack of scientific certainty is no legitimate excuse for hindering the development of technologies which could help avoid or minimize the possibly disastrous consequences of unchecked climate change and additional environmental stresses. Not surprisingly, proponents of geoengineering experiments – such as the Copenhagen Consensus Center, the Arctic Methane Emergency Group, and a number of geoengineering researchers (Blackstock et al. 2009) – implicitly prefer and act on the latter interpretation. And to conclude, this is also the sort of implicit interpretation that underpins the Old Massett Chief Councillor Ken Rea desperate defense of the Haida Gwaii combined ocean fertilization–salmon restoration project.

'Tipping Points', Surprise and Innovation

I believe there is a deeper underlying issue here, and that is how we sensibly should approach technological innovation in the Anthropocene, keeping in mind that complex systems entail a number of properties – previously described as connectedness, thresholds and surprise (Chapter 2). More precisely, the risk of transgressing dangerous climate thresholds, possibly regional nonlinear changes on the Arctic ice sheet with propagating consequences for global climate regulation, and the risks of irreversible changes in important ecosystems, are increasingly used to support the development (and at times also the deployment) of geoengineering technologies (Lynas 2011, Bipartisan Policy Center 2011, p. 3). The same properties can also result in considerably more cautious analysis of the expected benefits of experimenting with, and deploying geoengineering technologies (Steffen et al. 2011, Galaz 2012, Robock 2008).

This ambiguity is clearly visible as scientific advice moves into prominent international policy arenas. The ad hoc expert group created by the

Convention on Biological Diversity and in which I took part (mentioned earlier), reaches a number of conclusions that illustrate the indefinite role the existence of complex systems behavior, such as thresholds and connectedness, plays for actual policy choices. For example, one key conclusion is that '[t]he deployment of geoengineering techniques, if feasible and effective, could reduce the magnitude of climate change and its impacts on biodiversity. At the same time, most geoengineering techniques are likely to have unintended impacts on biodiversity, particularly when deployed at a climatically-significant scale, together with significant risks and uncertainties' (Williamson et al. 2012, p. 14).

Coping with scientific, and other forms of social, technological and environmental uncertainty has obviously always been a governance challenge (Koppenjan and Klijn 2004, Pierre and Peters 2005). The issue here is that the promotion of innovation of all sorts, often is identified as a fundamental strategy to stay ahead of change in complex systems behavior in the Anthropocene (Westley et al. 2012, Olsson and Galaz 2012). In more theoretical terms, novelty has always been a central element of resilience thinking (Folke et al. 2010). Frances Westley, sociologist at the Waterloo Institute for Social Innovation and Resilience (Canada) formulates it nicely:

> With the earth and its ecological systems pushed close to planetary boundaries, we need innovative solutions that take into account the complexity of the problems and then foster solutions that permit our systems to learn, adapt, and occasionally transform without collapsing. More important, we need to build the capacity to find such solutions over and over again. (Westley 2013, p. 6)

Thomas Homer-Dixon raises the same point in his inspiring book *The Ingenuity Gap* (2002). Starting from a complex systems perspective, Homer-Dixon notes that the 'central feature of societies that adapt well is their ability to produce and deliver sufficient ingenuity to meet the demands placed on them by worsening environmental problems' (Homer-Dixon 2002, p. 21). A failure to innovate creates a widening 'ingenuity gap' with possibly severe social and environmental problems as societies try to grapple with unfolding changes, surprises and shocks.

This poses a fundamental conundrum that few scholars have explored in any detail (Galaz 2012). On the one hand, the risks entailed with the transgression of large-scale critical biophysical and ecological thresholds require societies to invest in increased innovative capabilities. Advancing experiments, 'remixing', 'hacking', and making creative use of ideas regardless of their origins, become quintessential from this perspective. At the same time, the interconnected features of biophysical systems and Earth system functions (see Chapter 2), as well as the possibility for unwanted surprises also imply a call for caution, especially as the systems to be experimented with, for example, a lake, a coastal zone, a forest can contain threshold dynamics. Robock and colleagues made a very important observation in 2010, which relate to the challenges posed by Earth system complexity (p. 531):

> [t]he signal of small injections would be indistinguishable from the noise of weather and climate variations. The only way to separate the signal from noise is to get a large signal from a large forcing, maintained for a substantial period.

This would make experimentation very difficult to separate from an actual large-scale deployment (see MacMynowski et al. 2011 for a different opinion).

C.S. 'Buzz' Holling – the founding father of resilience thinking – coined an interesting term of relevance here: fail-safe experimentation (Holling 2004). In short, as we enter a turbulent future, 'it is important to encourage experiments that have a low cost of failure to individuals, the environment, and careers, because many of these experiments will fail' (ibid).

Obviously this begs the question how to decide which of the experiments are to be perceived as 'safe', at what spatial and temporal scale, and for whom. For example, should we really deploy a few hundred marine cloud-brightening ships in the Arctic, for say 5–10 years, just to gather data on whether these are able to cool down the region, and help some of the ice sheet to recuperate? And maybe the Haida Gwaii experiment makes sense as it could provide multiple benefits to both local communities and the global climate, if proven successful, extended and upscaled? And if these sorts of experiments are not allowed, how are we to maintain our innovative capacities as we enter a turbulent and uncertain future where humans are placed as the stewards of the Earth system?

These are not only provocative questions, but also issues which will require innovative thinking about possible institutional options. And as noted, one of the critical pillars of environmental policy – the precautionary principle – is certainly not going to be of much help in this endeavor.

Planet Engineers

As already noted, geoengineering is far from a new topic. The main point here is that the geoengineering debate marks what could be only the beginning of several evolving challenging social and political debates of the role of technologies in an era of rapid environmental change, and a deeper understanding of connected risks embedded in systems with complex system properties. Why? Consider the following examples.

De-extinction: On 15 March 2013, the prestigious journal *National Geographic* hosted a so-called TEDx-event, an independently organized and pop-science event organized in the same way as the original TED-talks: captivating short talks about either deep societal problems, or scientific insights, or both. The issue was 'de-extinction', that is the evolving possibility to bring extinct species back to life through genetic technology. As Stewart Brand – multi-entrepreneur, author and president of the *Long Now Foundation* that hosts the de-extinction initiative 'Revive and Restore' – notes, many extinct species might be extinct bodily, but not genetically. That is, it is technologically possible to revive extinct species through cloning and by using DNA samples obtained from museum specimens, frozen tissue samples, and fossils. A few early de-extinction experiments are already under way, including the passenger pigeon (extinct in the early twentieth century) and the woolly mammoth (extinct about 4000 years ago). These sorts of projects are not only driven by human curiosity, but seemingly also concerns of rapid environmental degradation. Or in the words of Brand: 'Why do we take enormous trouble to protect endangered species? The same reasons will apply to species brought back from extinction: to preserve biodiversity, to restore diminished ecosystems, to advance the science of preventing extinctions, and to undo harm that humans have caused in the past' (from Brand 2013). Opponents on the other hand, argue that not only are 'de-extinction' scientists underestimating the challenges of correctly transforming DNA into living species, but also that the habitats of extinct species have either been drastically transformed or essentially disappeared. In the words of biology professor David Ehrenfeld: 'Right now, de-extinction is just an interesting idea, what we might call recreational conservation. Pursue the dream if you like, but ease off the hype' (Ehrenfeld 2013).

Synthetic biology and conservation: One month later, the Wildlife Conservation Society and the Nature Conservancy hosted a workshop at Clare College in Cambridge, exploring the interface between synthetic biology and conservation. Synthetic biology is in many ways a much more far-reaching technology compared to conventional genetic technology. The reason is that synthetic biology allows scientists to put together entirely new combinations of synthetic DNA from raw materials. One of the most well-known examples is American biologist and world famous entrepreneur Craig Venter and his team's successful attempt (in 2010) to create artificial life, or more precisely, an almost completely synthetic version of the bacterium *Mycoplasma mycoides*. Synthetic biology has an often-acclaimed enormous potential, ranging all the way from designed enzyme or bacteria able to produce a new generation of biofuels, to engineered crops, and more rapid restoration of damaged or destroyed ecosystems (Redford et al. 2013). Interestingly enough, the framing report to the meeting in Cambridge between synthetic biologists and conservation specialists, starts with an exploration of the Anthropocene and the risk of 'abrupt and irreversible state shift in the Earth's biosphere' (p. 2).

Unconventional coral reef protection: As already explored in the previous chapter (Chapter 3), the world's oceans are the scenery for multiple interacting global environmental stresses, such as ocean acidification, loss of marine biodiversity, and eutrophication (Nyström et al. 2012). The perceived dire straits and the closing 'window of opportunity' to maintain anthropogenic global warming below +2°C in the next few decades, has triggered arduous scientific debates on whether time has come to prepare for 'unconventional' and 'non-passive' ocean conservation approaches. In 2012, Greg H. Rau and colleagues suggested that current 'policy statements will prove inadequate or ineffective' for oceans as we enter an era of rapid environmental stresses, and that 'a much broader evaluation of marine management and mitigation options must now be seriously considered'. These imply large-scale technological interventions, such as the creation of protective shields, ocean fertilization, artificial reefs and genebanks to help restore coral reef ecosystems in the face of ocean acidification and climate change (Rau et al. 2012). One interesting observation here is that Rau and colleagues conclude their argument for unconventional and non-passive strategies by referring to the precautionary principle as stated by the Convention on Biological Diversity in 1992: 'Where there are threats of serious or irreversible damage, lack of full scientific certainty shall not be used as a reason for postponing cost-effective measures to prevent environmental degradation' (Rau et al. 2012, p. 4).

Do these examples seem extreme? Consider then instead this last example – the emerging debate in the conservation community between conservation scientists, ecologists and managers. The issue is the idea of 'assisted migration' – the intentional relocation of species as a means to protect them from extinction, a risk created by unfolding environmental changes such as climate change. Proposals range from building cocoons to assist the migration of the quino checkerspot butterfly in California, to the relocation of polar bears from southern regions to cooler regions that have sufficient sea ice cover (Marris 2008, Struzik 2013). For some scholars, assisted migration is likely to be the only way to protect species (and hence biodiversity) unable to migrate or adapt fast enough to keep up with the rates of environmental change (Hoegh-Guldberg et al. 2008, Derocher et al. 2013). For others, assisted migration entails similar risks as those created by invasive alien species, and this policy option should therefore be either effectively bolted, or at least strictly regulated (McLachlan et al. 2007, see Hewitt et al. 2011 and Heller and Zavaleta 2009 for syntheses).

Table 5.1   Unconventional technological proposals to address planetary boundaries

Planetary boundary	Suggested intervention	Reference/s
Climate change	Solar radiation management and carbon dioxide removal	This chapter
Atmospheric aerosol loading	Solar radiation management	This chapter
Stratospheric ozone depletion	Creation of global or regional artificial ozone layer by microwave discharges	Gurevic and Borisov (1995)
Ocean acidification	Protective shields, creation of artificial reefs, liming the oceans, gene banks	This chapter, Rau (2011)
Biodiversity loss	De-extinction; assisted migration; combining synthetic biology and conservation; underwater vacuum cleaners or 'supersuckers' to actively remove invasive algae; restoring marine oxygen depleted 'dead' zones by continuously pumping oxygen rich water to the deep water	This chapter, Redford et al. (2013), Stigebrandt (2012)
Chemical pollution	None found	
Global freshwater use	Global artificial photosynthesis	Faunce (2012)
Nitrogen and phosphorous cycle	None found	
Land use change	Global artificial photosynthesis	Faunce (2012)

The legal possibilities and obstacles remain to be explored (Marris 2008), but allegedly '[t]his strategy flies in the face of conventional conservation approaches' (Hoegh-Guldberg et al. 2008, p. 345).

I could list more examples of course. Table 5.1 summarizes a range of non-conventional proposed technological interventions as a means to reduce pressure on 'planetary boundaries'. The list is not complete in any way, and we could debate at length over the term 'non-conventional'. My main point is the following: geoengineering and the types of intricate political questions and governance challenges these technologies raise, are likely to be only one type of emerging controversial technologies and proposals prompted by attempts to cope and adapt to rapid and possibly nonlinear environmental change. Technological suggestions that currently are at the margins of the debate, could in times of perceived urgency and crisis, enter central policy discussions.

These technologies and proposals raise similar governance questions explored earlier for geoengineering: should experiments be allowed, and if so, for how long, and at what scale? Does the possible transgression of dangerous biophysical and ecological thresholds implicate more aggressive investments in innovation, or a more careful approach? Which organization or legislative body is allowed to make that choice, and how should we interpret the 'precautionary principle' in this context? Are the risks entailed with experimentation or deployment likely to be larger, or smaller than those projected in a business-as-usual scenario? Which scientific body has the legitimacy and competence to make such an assessment of likely trade-offs? And what would an effective polycentric institutional architecture look like if it were able to address not only financial issues, patent conflicts and questions about experimentation, but also tough geopolitical dimensions? These questions neatly illustrate the difficult governance and political challenges posed by the 'Anthropocene Gap'.

A NOTE ON CONVERGING TECHNOLOGIES

I'm aware that the list above is likely to provoke some of the readers to consider it as mere science fiction, or weird outliers with little, or no relevance whatsoever in discussions about the future of governance in the Anthropocene. That perception, I believe, would be truly incorrect. The rate of technological development is not only fast, but in many aspects also nonlinear. As Béla Nagy and colleagues recently showed, not only computing hardware, but also a range of technologies seem to follow Moore's law of rapid nonlinear progress (Nagy et al. 2013). What's more, many of the examples elaborated – such as geoengineering and deextinction – above illustrate the converging nature of technologies, as they coevolve and drive further innovation (Roco 2008, Arthur 2009, Kelly 2010).

Geoengineering technologies as an example, include both proposals to inject nano-particles in the stratosphere (Keith 2010), as well as suggestions to design biotechnologically modified crops to enhance their sun reflecting (in other words, 'albedo') and hence climate cooling properties (Rigwell et al. 2009). Suggestions to advance and support 'global artificial photosynthesis' technologies that would reduce stresses on many 'planetary boundaries' including climate, water and land use change, build extensively on advances in nanotechnology (Faunce 2012). What's more, a sensible discussion about the impacts of these technologies on the Earth system would be impossible without the rapid advancements in computer processing powers that underpin today's highly complex climate models. De-extinction technology and supporter networks seem to effectively tap into not only genetic engineering technologies and potentially also synthetic biology, but also into the potential of the Internet through tutorial videos, collaborative wikis, and blogs. Synthetic biologists not only benefit from automated assembly technologies (in other words, robotics) and advanced computer models, but also attempt to fund their experiments through online micro-funding sites (for example, mycroryza.com). Processes of 'collective intelligence' such as those explored in Chapter 4, are likely to be abundant as actor networks evolve and overlap in their attempts to push disruptive innovation further (Moore and Westley 2011).

Brian Arthur's observation that technology matures, diversifies, and scales at an accelerating pace (Arthur 2009) is an important reminder of why these issues are far from negligible if we are interested in the challenges to global environmental governance posed by interplaying environmental and technological change.

6. Financial markets, robots and ecosystems

Here's a story you've probably never heard of before. Anne Hathaway is a young 30-something American actress, known from movies such as *The Devil Wears Prada* (2006), and the Batman movie *The Dark Knight Rises* (2012). One interesting thing with Ms. Hathaway (at least for the purpose of this book) is that every time she makes the headlines, the stock of Warren Buffett's Berkshire–Hathaway goes up. In 2010 when Ms. Hathaway co-hosted the Academy Awards, the Berkshire–Hathaway stock made two unexpected upward jumps. First, on the Friday before the Oscars, Berkshire shares rose over 2 percent. On the Monday following the awards, they rose again by over 2.9 percent (Mirvich 2011, Stodola 2011).

News analytics via automatized monitoring of 'chatter' on the Internet is making fast progress in stock markets. It's not a far-fetched guess that this bizarre correlation between the value of a company's stock, and the actions of a young actress, is created by recent and rapid advances in news analytics and automatized trade. The mechanism *could* be the following: A computer monitoring news feeds (in a similar way as done by GPHIN for epidemics as explored in Chapter 4) finds an unusual number of positive Hathaway mentions (in other words, the actress), and rapidly executes buying orders in shares of the company. As stocks go up, the ultrafast buying computer has executed a successful 'momentum trade' as other slower traders aim to buy the now 'hot' share.

How is this story related to the issues outlined in previous chapters of the book? As I intend to discuss, algorithmic trade – that is, computerized, microsecond-fast trade – with commodities such as sugar and wheat, is an example of how emerging technologies create novel, complex and intriguing governance challenges in the Anthropocene. In essence, the increased interconnection between financial and commodity markets and associated technological advances, display many of the intriguing features elaborated in this book: complexity, connectivity, thresholds, surprise and cascades.[45]

The chapter is structured as follows. First, I will introduce the notion of 'robot-trade', 'computerized trade' or 'algorithmic trade' (henceforth AT), and propose a simple definition of the concept. Second, I explain how algorithmic trade and an increased interlinkage between financial markets and commodity trade pose new and poorly understood social–ecological challenges. Third, I map out the most important governance responses, some critical tensions, and draw out some more general theoretical insights based on this case.

RISE OF THE MACHINE

Computer based trading with commodities has a long history, with some early examples of the creation of electronic marketing systems for cotton in the US in the mid-1970s (Lindsey et al. 1990). As the capacities of computer algorithms to process increasing amounts of information, and extremely rapid (in other words, microsecond) and complex trading patterns have increased, so has their progression in a wide set of markets such as those for bonds, stocks and foreign exchange markets.

Algorithmic trade[46] involves several types of automated trading strategies, each with different ways of making profit with minimal human intervention. In short, AT has four main comparative strengths compared to conventional trade (based on McGowan 2010, p. 3, Gomber et al. 2011, p. 21). First, the ability to split up large orders, and execute them in such small portions over time, so that market impacts (for example, increases in the price) are minimized. Second, the ability to text-mine different sources of information for 'weak signals' – such as the movement of interest rates, very small economic fluctuations, sentiments in financial news and social media such as the micro-blog *Twitter* – and rapidly execute trade before anyone else in the market is aware of them. Third, algorithms benefit from identifying and rapidly trading on differences in prices between exchanges and platforms, known as 'arbitraging between spreads'. Fourth, algorithms are able to place 'flash orders' – that is trade based on access to market information for a matter of milliseconds, which is not yet public. While the benefits can be extremely small for each individual trade (pennies of a US dollar for example), this sort of trade can be executed millions of times a day, thereby generating large gains in aggregate. For example, despite the economic recession, algorithmic trade is seen as one of the most lucrative businesses on Wall Street, and is estimated to generate approximately 15–25 billion US dollars in revenue yearly (McGowan 2010, p. 3).

It should be noted that AT is far from an outlier phenomena. Large market players such as hedge, pension and mutual funds have been reported to increasingly execute large trades through algorithmic trading. Early estimates show that algorithmic trading has increased from near zero in the mid-1990s, to as much as 70 percent of the trading volume in the United States (Hendershott et al. 2011, p. 1), 30 percent in the UK, and between 30 and 50 percent in the European equities markets (Government Office for Science 2011). According to Johnson and colleagues (2013) the impacts of algorithmic trade is much more profound than can be captured by simple percentage estimates. On the contrary, they suggest that the emerging 'ecology' of trading technologies and algorithms has induced an 'abrupt transition to a new all-machine phase' (Johnson et al. 2013, p. 1) implying a drastic systemic transformation in the speed and nature of financial trade displaying new and unprecedented financial phenomena.

Historically, AT has only been a concern in financial markets, such as US-based equity and foreign exchange markets. This is changing rapidly as AT technologies make their entrance into commodity markets (Mitra et al. 2011). For example, UBS Investment Bank already in 2007 announced the launch of a Commodities Portfolio Algorithmic Strategy System that allows for automated trade on 19 commodity futures markets. Recently developed news analytics software, such as the Reuters NewsScope Sentiment Engine, allow for sentiment analysis in milliseconds for news on 40 commodity and energy assets in addition to over 10,000 companies. Major commodity exchanges such as NYSE Liffe, now also offer co-location services, that is, they provide all the physical infrastructure needed to collect, analyse and store data close to their servers. This allows algorithmic engines to operate as fast as possible in Chicago-based commodity markets – the largest commodity market in the world.

How does this technological development affect the biosphere or ecosystems? And in what ways does this provide interesting insights to the underlying theme of this book? Before answering these questions, two additional trends need to be understood: the rapid increase of financial derivatives in commodity markets and the increased interconnectedness between financial, and commodity markets.

A Rapid Increase of Commodity Derivatives

One important trend in commodity markets globally, is the rapid increase of trade with so-called *commodity derivatives* such as futures, options, commodity indexes, exchange tradable funds, and certificates. Commodity derivatives are (bluntly put) financial instruments created to deal with the economic risks created by price fluctuations. Price fluctuations are problematic for both buyers and sellers in a market: the first risks losing valuable income if prices drop unexpectedly, and the latter risks seeing their profits drop rapidly if costs for raw materials increase unexpectedly. A commodity futures contract (one of several forms of commodity derivatives) is an agreement to trade the physical commodity in the future. It specifies the price and quantity of the commodity, and the time and place where the physical trade will be carried out. This means that traders wishing to buy or sell a physical commodity at some time in the future can use futures contracts to insure themselves against price fluctuations (Geman 2005).

Commodity derivatives are in no way a novel phenomenon. Some forms of derivatives were used in seventeenth century Japan for rice, and more formalized trade of futures contracts for commodities began in Chicago in the 1840s (Lambert 2011). The commodities for which derivatives are traded today include canola, cocoa, coffee, corn, cotton, feeder cattle, lean hogs, live cattle, lumber, oats, orange juice, rice, soybeans, soybean oil, sugar, and wheat.

What is interesting in this context is the extremely rapid increase of trade with options and futures on many commodities such as oil, grains, soybeans, cocoa and coffee. For example, the value of commodity futures contracts in the US doubled to an estimated US$400 billion between the years 2005 and 2008 (Clapp 2009, p.1187). Investments in commodity index funds – that is, funds containing investments in multiple commodity derivatives – increased from $13 billion in 2003 to $260 billion by 2008. In the period 2005–2008 commodity index funds alone drove as much as 300 billion dollars into US commodity futures markets (Girardi 2012, p.86). Agricultural commodities – which have tangible links to multiple ecosystem services – typically account for around 30 percent of the commodities in these index funds (Clapp 2009, p.1187).

The reasons for these observed rapid increases in trade with commodity derivatives are contested, but often viewed as an effect of financial deregulation through, for example, the Commodity Futures Modernization Act (CFMA) of 2000 (United Nations Conference on Trade and Development 2009, pp.76–77), increased global demand, and the emergence of new actors in commodity markets (see below).

The 'Financialization' of Commodity Trade

Not only has the number of financial instruments for commodities increased, so has the number of heavyweight actors in commodity trading. More precisely, large financial investors – amongst others sovereign wealth funds, pension funds, hedge funds, and university endowments – have moved into commodity futures markets for commodities such as wheat, sugar, and cocoa (United Nations Conference on Trade and Development 2011, p.18).

This progression is the result of attempts by financial investors to diversify their investments portfolios in the face of global economic uncertainty (Gorton and Rouwenhorst 2006, TheCityUK 2011, p. 11). In short, investments in commodity futures help financial investors to create a portfolio consisting of a broader set of assets, which can act as a buffer from turbulence in, for example, bonds and currencies.

Hence apart from providing means of insurance for the participants in the markets for the physical commodities, derivatives markets also provide the opportunity for a wider set of financial actors to gain profit by betting (or speculating) on the development of the prices of commodities. This means that financial actors in commodity derivatives markets are not necessarily interested in the underlying commodity (say, sugar and cacao), but rather in the financial instrument.

As the US equity market bubble burst in 2000, investments in commodities arose rapidly throughout the world as a way of reducing financial risks against inflation and changes in the exchange rate of the US dollar (Girardi 2012, UNCTD 2011, p. 13). However, these rapid flows of money into commodity markets seem to have slowed down in the few last years due to (amongst other things) the European debt crisis (Terazono 2012).

It is interesting to note how these three trends intermingle: a rapid increase in financial instruments for commodities develop (triggered partly by legislative changes and a global economic crisis), and co-evolve with the movement of large financial actors into these markets, and associated novel trading technologies. Let us now explore why this matters for the analysis of governance in the Anthropocene.

PRICE VOLATILITY, ECOSYSTEMS AND PEOPLE

There have been few reasons for ecologists and global environmental change scholars to engage in-depth with the behavior of financial markets in the past. While recent studies explore the similarities between financial and ecological systems as complex adaptive systems, and their disposition to abrupt change and state shifts (for example, May et al. 2008, Scheffer et al. 2012), few studies have explored the linkages between the two. A few exceptions are emerging and there is strong criticism against 'the financialisation of environmental conservation' (Sullivan 2012), the 'commodification' of ecosystem services through Payments for Ecosystem Services (Kosoy and Corbera 2010), and the implications of the expropriation of land in the global South for environmental conservation reasons (so-called 'green grabbing') (Fairhead et al. 2012). The absence of studies which connect financial instruments, and the behavior of financial and commodity markets with ecosystem change is surprising considering that market information (such as experienced and projected changes in prices including volatility) is known to have downstream impacts as companies, farmers, fishermen and diverse production cooperatives adapt to economic opportunities and risks (Eakin and Wehbe 2008, Deutsch et al. 2007, Lambin et al. 2001). In addition, the production of commodities has tangible impacts on ecosystem goods and services, including biodiversity (Fischer et al. 2006). The first step in my mind is to identify key uncertainties, and map out current controversies.

Does the Rapid Progression of Financial Actors into Commodity Markets Matter?

One common view is that the progression of financial actors into commodity markets can improve the functioning of markets by increasing the amount of contracts available for insurance, and by adding information to the price formation process (Grossman 1977). Nevertheless, international policy-makers and non-governmental actors have become increasingly concerned about the possible unintended impacts of the evolving interconnection between financial and commodities markets. The main concern is the possibility for financially induced price instabilities as volatilities – that is, unexpected and rapid changes in prices – to propagate through different asset classes.[47] Bluntly put, there are concerns that an increased interconnection could lead volatilities in, for example, foreign exchanges and derivate markets, to cascade and disturb the price mechanism of key food commodities such as wheat, maize and soybeans (United Nations Conference on Trade and Development 2011, p. 11, see also Bicchetti and Maystre 2012, Harmon et al. 2010). In this scenario, novel financial instruments for commodities would not reduce, but rather increase risk due to increased connectedness.

While price volatility has always existed in commodity markets (Gilbert and Morgan 2010), the concern is that the progression of financial actors into commodity markets could cause commodity prices to be determined by not only supply and demand of the underlying commodity, but also by financial factors and expectations of other investors' behavior (United Nations Conference on Trade and Development 2011, p. 25).

This has raised highly contested concerns about the risks of herd behavior, speculative bubbles, and unexpectedly high price volatility in commodity markets as financial actors try to predict the behavior of multiple markets, and of each other (United Nations Conference on Trade and Development 2011, p. 55, see also Girardi 2012).

In other words, this is a truly interconnected complex adaptive system with poorly understood feedbacks between actors (for example, traders, producers and consumers), technology (for example, trading technologies) and downstream human–environmental interactions (for example, production of commodities all over the world and their associated ecosystem impacts).

The probability for financially induced price disturbances of this sort is suggested to be higher for trade with 'soft' commodities such as coffee, cocoa, sugar, corn, wheat, and soybean. The underlying reason is that financial players trade comparatively very large volumes, and can shift rapidly from one market to another, thereby causing 'substantial volatility and price distortions' (United Nations Conference on Trade and Development 2011, p. 43, see also Borin and Di Nino 2012).

The implications of increased price instabilities should not be underestimated. Price volatilities for commodities are known to have severe negative macroeconomic and food security impacts (for example, United Nations Conference on Trade and Development 2011, FAO 2011, de Schutter 2010). Recently, the World Economic Forum put extreme volatility in energy and agricultural prices as one of the top economic risks for the future, fearing that these might trigger conflict between social groups and spur increased geopolitical tensions (World Economic Forum 2013).

International policy concerns escalated after the 2008 'food crisis' which showed an unusually high volatility for food commodities, a phenomena proposed to be linked to the 'speculative' behavior of financial investors (de Schutter 2010). This, however, is a highly debated conclusion. BüyükŞahin and colleagues (2010), Gilbert and Morgan (2010), Irwin and Sanders (2011) argue that claims of increased price instabilities due to tighter cross-market linkages, is premature considering the lack of empirical data, and considerably less straightforward than portrayed.

An interesting note is that prices as well as price volatility for food commodities are likely to continue to increase as we enter the Anthropocene, due to continued population and economic growth, changing diets in emerging economies, as well as a projected increased use of biofuels, rising oil prices and possible climate change impacts on agriculture (FAO 2011, p. 11, see also OECD–FAO 2011).

The Debate

Algorithmic trade is drawing considerable attention from the policy community. However, the impacts of AT on price dynamics is highly contested. Two main positions can be identified in the debate. One is that AT reduces the costs of trading, enables risk sharing, and improves the ability of assets to be sold without effect on the market (known as *liquidity*, see Hendershott et al. 2011, p. 1).

The second position instead holds that AT could contribute to sudden large price swings and market instability, triggered by unforeseen feedback-loops, errant algorithms and/or incorrect input data, or manipulative uses of algorithms to obfuscate actual prices (Johnson et al. 2013, GOS 2011, pp. 13, 36, Lenglet 2011, p. 59, Chaboud et al. 2009). This element of 'surprise' created by complexity explored in the opening chapters can be exemplified with the 2010 'Flashcrash'. In the course of about 30 minutes, US stock market indices, stock-index futures, options, and exchange-traded funds experienced a sudden price drop of more than 5 percent, followed by a rapid rebound (US Commodity Futures Trading Commission and US Securities and Exchange Commission 2010). Similar unexpected and very rapid volatility in prices for oil and gas futures have been observed the last few years (GOS 2011, p. 36). One of the most extreme examples must be the ultra rapid $136 billion loss (and recovery) of the value of the stock market index S&P500 in April 2013. In this case, the acclaimed reason was a hijacked *Twitter* account of the major news company Associated Press, sending out clearly inaccurate messages of an attack on the White House, which triggered rapid sales on the market.

The debate about the benefits and risks associated with algorithmic trading has until now had financial markets as an initial focus. But the debate is gaining ground in commodity markets as well. As mentioned, price volatility has always been a general feature of commodity markets (Gilbert and Morgan 2010), but the progress of algorithmic trade has been claimed to introduce additional disturbances. The commodity industry has raised concerns that algorithmic trade is contributing to 'extreme' price volatility for commodities such as sugar and cotton, as it allows large actors to move in and out from commodity markets exceedingly rapidly.

In November 2010 for example, raw sugar futures traded at the US Intercontinental Exchange (ICE) 'suffered their biggest one-day sell-off in 30 years', followed by more than 20 percent drop in sugar prices over two days (Blas 2011, Blas 2012). This claim is highly debated, and counterclaims state that algorithmic trade is only a small fraction of trade in commodities, and instead reduce volatility (Rampton 2011). Despite existing controversies, the US-based Intercontinental Exchange (ICE) Futures amended its rules in 2011 to be able to adjust trade prices to be more able to deal with volatility in its coffee, cotton, sugar and cocoa futures markets (Nicholson 2011). These are illustrative examples of the sort of events that policy-makers are trying to grasp, and regulate (Gomber et al. 2011, Lenglet 2011).

The temporal dynamics of this issue are multifaceted: By that I mean that different factors interact at very different temporal dimensions. For example, 'speed-traders' and their algorithms operate at the scale of milliseconds (1/1000 of a second); price fluctuations play out over days and months; and large financial index investors, local producers and production cooperatives operate over temporal scales of years. How tightly these cycles interlock – and the possibility for surprises to propagate – is still debated. Interestingly, it seems like the 'telecoupling' created by technologies, actors and financial markets, is able to transfer disturbances at these different temporal scales together. Neil Johnson and colleagues (2013) recently presented results from millisecond scale trade data in the global financial market, and show not only an increase of extremely rapid crashes and rebounds (so-called 'ultrafast extreme events', or UEEs with a duration of less than 1500 milliseconds), but also that these micro-crashes are seemingly linked to larger system instabilities such as the global financial crash in 2008. In their own words:

> Figure 1C therefore suggests that there may indeed be a degree of causality between propagating cascades of UEEs and subsequent global instability, despite the huge difference in their respective timescales. [. . .] Fig. 1C demonstrates a coupling between extreme market behaviours below the human response time and slower global instabilities above it, and shows how machine and human worlds can become entwined across timescales from milliseconds to months.

The conclusion is bound to be debated, but one observation is critical for the ambitions of this book: algorithmic trade is rapidly progressing in global commodity markets. The map in Figure 6.1 (based on the work presented in Galaz et al. 2013) is an early assessment of where in the world algorithmic trade with commodities has emerged. The next section explores why that matters.

Exploring the Connections

What are the linkages between financial and commodity markets, and ecosystems and ecosystem services? The academic literature is highly sparse in this regard, but there is one relevant common issue in the debate, which also has tangible implications on ecosystems: price volatility. Rapid fluctuations in markets have known impacts on how ecosystem stewards such as farmers and fishermen and different forms of cooperatives, manage ecosystem goods and services on land and in seascapes.

Complex causal relationships is a critical issue here: The drivers of land-use change are highly intricate (Lambin et al. 2001), but there is one key determinant: economic information (Lambin 2005). As farmers respond

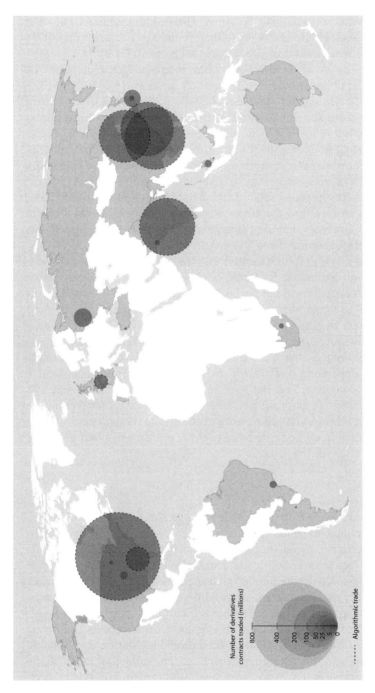

Notes: The map shows the world's 20 largest commodity derivatives markets. Data on trade volume from World Federation of Exchanges (2011). *Improving Structure, Promoting Quality – Annual Report 2011.* World Federation of Exchanges, Paris. Information about indications of algorithmic trade (. .) is based on online searches in financial news archives. See Galaz et al. (2013) for details.

Figure 6.1   The world's 20 largest commodity markets, and indications of algorithmic trade

to price information including volatility, the resulting actions alter the local biophysical environment by either reducing, or increasing their resilience to future environmental shocks (Adger et al. 2009).

Price volatility in coffee markets is one interesting case in point. In response to the collapse of the coffee market in the end of the 1990s due to declining public support and environmental stresses, coffee farmers in Mexico undertook a range of actions waiting for better future prices. Examples include allowing their coffee plantations to return to forest cover; converting coffee to sugar cane or pasture; selling coffee orchards to urban developers; or experimenting with other cash crops. These shifts in land use have been suggested to give broader-scale biodiversity and ecosystem services impacts, such of the viability of cloud forest habitats, net release of carbon into the atmosphere, and changes in local climatic controls such as winter humidity (Eakin and Wehbe 2008, see also Eakin et al. 2009 for a case study from Vietnam).

It should be noted that the problems created by unusually high price volatility is not only a dilemma for farmers in the global south. Concerns were widespread amongst farmers in the United States in 2008, as abnormal volatility seemed to undermine the effectiveness of existing commodity trade derivatives such as futures and options, making known hedging instruments not only more expensive, but also less reliable (Henriques 2008).

Yet local ecosystem management responses to price volatility are likely to be diverse depending on the temporal scale of volatility, the particular commodity of interest, and the specific sociopolitical, institutional and economical context (Lambin et al. 2001). The interlinkages between financial markets, commodity trade, price dynamics and ecosystem change hence remain complex, contested and poorly understood. What are some possible scenarios for the future?

Two Scenarios

In (Galaz et al. 2013) we explore two possible scenarios, each with very different outcomes depending on how uncertainties are interpreted. In a best-case scenario, novel financial instruments and trading algorithms effectively reduce price volatility and transaction costs. More robust price setting dynamics would then underpin novel flows of economic resources into commodity markets, which eventually would increase necessary production and investments in, for example, agriculture. Remaining price volatilities could be dealt with through new insurance instruments, and/or by the diversifying land use. Depending on national legislation and policies, as well as existing public–private partnerships, incoming funding could be used to combine sustainable production with stewardship of ecosystems.

While this might sound overoptimistic, Koh and Wilcove (2007) explore similar issues for oil-palm agriculture. They propose that incoming revenues from increasing prices in world markets, could be used to create networks of privately owned nature reserves, combined with the continued development of 'self-sufficient villages, providing not only employment, but also housing, basic amenities such as water and electricity, and infrastructure including roads, medical care and schools for the families of their employees' (Koh and Wilcove 2007, p. 994).

A similar scenario could be outlined for additional agricultural commodities and agro-ecosystems. Financial instruments such as forward contracts have historically provided farmers with economic resources as the same time as securing buyers with a predictable supply of raw materials (De Schutter 2010, p. 9). In addition, land investments following from an increased interest in commodity markets by financial actors, could through 'responsible agricultural investment' at best drive increased productivity, and technological investments, and contribute to poverty reduction (FAO 2010).

The creation of new financial instruments could also transform economic incentives and support effective stewardship of resources currently outside of commodity futures markets. Recently for example, the creation of financial instruments for fisheries (which currently lacks a market for futures) has been suggested as a means of preventing stock depletion (Dalton 2005). Others have argued that new financial instruments could lead to more effective conservation of endangered species, as it would stimulate private actors to take preventive action through, for example, private conservation efforts (see Mandel et al. 2010).

Again, this scenario is heavily dependent on the ability of market actors and policy-makers to create appropriate legislative frameworks, national programs, financial mechanisms, and associated monitoring and implementation capacities, to secure that incoming funding is channeled in ways that support sustainable ecosystem stewardship.

There is also a pessimistic scenario of course. This scenario would view the rapid progression of commodity derivatives, the entrance of large and in economic terms powerful financial actors into commodity markets, and novel trading technologies, as an initiator of novel disturbances such as unforeseen cascading price volatilities. This would undermine necessary on-the-ground technological investments, and steer producers away from ecosystem stewardship.

Several mechanisms are possible here. Firstly, increasing price volatility could deter producers such as farmers and farmers' organizations from making investments in sustainable production methods. Intra-annual volatility for example, could force producers to sell when prices are low, hence undermining needed longer-term investments. Inter-annual volatility has similar impacts as producers struggle to generate stable high returns, access credits and meet fulfillments as defined in contracts (FAO 2011, p.37, FAO 2011b, p.12, El-Dukheri et al. 2011). Second, upwards price volatility could lead to spillover social–ecological effects as social actors shift their production or extraction of ecosystems goods and services. For example, people who depend on forests for their food security are often very vulnerable to higher food prices. Rapid increases in prices force these communities to collect more out of the forests with 'direct impact on forest quality, conservation and the survival of key forest species' (FAO 2011, p.16). In fisheries, price volatility has known negative effects on inter-temporal resource allocation decisions, for example the financing of new fishing gear or processing facilities (Sethi 2010). Third, as novel financial actors drive commodity future prices upward, the expansion of production beyond changes in supply–demand fundamentals is a theoretical possibility in the form of a 'speculative bubble'. Emerging international carbon markets have, as an illustration of this mechanism, been suggested to already be driving large-scale forest carbon offset investments in the global South, with controversial social and ecological implications (Fairhead et al. 2012). The UN Special Rapporteur on the Right to Food, Olivier de Schutter, argues that increases in price and volatility in food commodities is the result of a 'speculative bubble' that can be traced to herding behavior produced by commodity indexes. In one report presented in 2010, de Schutter argues that commodity index funds has produced a 'vicious circle of prices spiraling upward: the increased prices for futures initially led to small price increases on spot markets; sellers delayed sales in anticipation of more price increases; and buyers increased their purchases to put in stock for fear of even greater future price increases' (De Schutter 2010, p.4).

INSTITUTIONAL FRAGMENTATION, UNCERTAINTY AND MISFITS

Which scenario is most likely to unfold? The outcome depends fundamentally on the results of ongoing political discussions and attempts for more effective steering. Current initiatives are currently assumed by multiple actors and in different ways, ranging from self-regulation by large banks, to attempts by regulators to slow down hyper-speed trading. Below I will briefly go through a selection of these initiatives, before embarking on a more analytical exploration of the issue.

Government actors: The most debated reforms stem from *government actors*, mainly in the United States and in the European Union. The top US commodity regulator, the Commodity Futures Trading Commission, is keeping a close eye on algorithmic trading after its investigation of the above mentioned 'Flash Crash' on 6 May 2010, when the Dow Jones Industrial Average dived about 1000 points, only to recover within minutes. This grasp for oversight should be viewed in the light of the so-called *2010 Dodd–Frank Wall Street Reform and Consumer Protection Act*, a legal modification which brings considerable changes in to the US financial infrastructure as a way to increase transparency and government supervision of financial instruments and markets (Snider 2011). Or as financial news agency Bloomberg puts it: 'The Dodd–Frank financial legislation may make it easier for the Commodity Futures Trading Commission to punish manipulation and disruptive trading in markets for commodities such as oil, wheat and natural gas' (Loder 2010). In 2012, the US Securities and Exchange Commission also voted for the incorporation of a new system that will allow for a better overview of trade data and hence help audit the extent of algorithmic trade (High Frequency Traders 2012).

Regulation of this sort is nevertheless highly conflictive in US politics, and has led to intense lobbying efforts by Wall Street banks such as Goldman Sachs attempting to influence the details of the rules (Snider 2011, p. 12). The attempt by the Commodity Futures Trading Commission to limit speculation in commodity markets by imposing 'position limits' – that is, limits to the number of contracts traders can hold in 28 commodities, including oil, cocoa and coffee – drew considerable criticism from Wall Street institutions and politicians. In late 2012, for example, prominent republican Congressmen publicly opposed position limits stating that 'we are very concerned, in the wake of the financial crisis, that CFTC staff are using limited resources to pursue ideological and political goals rather than using the resources allocated by Congress to carry out the direct requirements of the agency'. A few weeks earlier, a US federal court had rejected the proposal on 'position limits', inducing new rounds of still ongoing debates and legislative processes (Nasipour 2012).

The European Union is another battleground for this debate. France has eagerly pushed for restraints on what has been seen as speculators such as hedge funds making quick gains in commodities markets (EurActiv 2011). Former President Sarkozy of France summarized the country's position satirically at a press briefing for journalists in January, 2011: 'I will recommend a date for the publication of a study showing that speculation does not result in global price rises of raw materials: 1 April', Sarkozy reportedly said. The acclaimed 'financialization' of commodity markets has in the last few years reached the highest levels of political decision-making in the European Union: the European Commission, as well as the European Parliament (EurActiv 2011b, see also report by the European Commission 2011). The latter adopted a first draft legislation based on the Markets in Financial Instruments Derivative (MiFID) in 2012 which attempts to put limits on what is called 'harmful speculation' which is argued to fuel price volatility on global agricultural commodity markets. Additional legislation targeting algorithmic trade was also adopted by the Parliament in October 2012 (Mifid II) (Price 2011). These are only the beginning of what is likely to become long and multiple negotiations between the decision-making bodies within the EU, as well as with member states and interest groups.

Meanwhile in India in the beginning of 2013, the Forward Markets Commission (FMC) announced its plans to put in place guidelines to oversee potentially adverse effects of algorithmic trade in commodity exchanges (Sahgal 2013).

International Organizations: These national government initiatives also have international counterparts. Several UN organizations such as the Food and Agricultural Organization (FAO), the United Nations Conference in Trade and Development (UNCTAD), have organized workshops and produced important synthesis reports that bring together the state of science and existing policy concerns (for example, United Nations Conference on Trade and Development 2011, FAO 2011). In 2008 and in the aftermaths of the 2008 'food crises', the UN Secretary-General Ban Ki-moon created a High-Level Task Force on the Global Food Security Crisis. The focus so far has been to support coordinated action between a myriad of UN agencies, international financial institutions and national governments to secure both short-term and long-term responses. In the first case, the focus has been on immediate humanitarian responses to food crises, and in the latter case designing mechanisms to secure long-term support to smallholder farmer food production (UN High Level Task Force on the Global Food Security Crisis 2009).

Private sector initiatives: The private sector has also launched a number of initiatives as a response to claims about destructive speculation with commodities, and the risks involved with hyper-speed algorithmic trade. At the end of 2012, for example, Barclays – one of the largest banks in the world – announced that it was considering quitting trading of agricultural commodities due to perceived 'reputational risk'. Similar signals to pull out from investments in food commodities have been sent by German banks such as Deutsche Bank, Commerzbank and Landesbank Baden-Württemberg (Slater 2012). Commodity exchanges have also responded to the critique. In 2011 for example, the ICE Futures US (ICE.N) announced that it would increase its ability to adjust trade prices in so-called 'softs futures' as a means to tackle volatility in its coffee, cocoa, cotton and sugar markets (Nicholson 2011).

INSTITUTIONAL DIAGNOSTICS

The contested complexities involved in these issues should not be underestimated. It is interesting to note the similarities of governance challenges in comparison with the other cases explored in this book. More precisely, four (by now I assume) familiar issues in particular stand out as critical governance challenges.

The first issue is *institutional fragmentation*. It is clear that the currently perceived problems emerging from the ongoing integration between financial and commodity markets and associated technologies, lack a simple overarching international institutional infrastructure. This fragmentation is particularly clear at the international level, despite the fact that the issue clearly is international due to the interlinked nature of global finance and trade. Current responses at the international level instead seem to focus on the creation of international polycentric coordination (see Chapter 3), with a strong emphasis on information sharing; pooling of funding for both long-term investments and short-term interventions; and internationally supported state action. Institutional reforms are assumed at national and supra-national levels exemplified by recent legislative changes in the United States, and ongoing legislative work in the European Union. Initiatives address both 'speculation' with commodities, as well as the risks entailed with hyper-speed trading technologies. But as for many similar issues explored in this book, responsibilities at the international level are highly dispersed, and require both international and national coordination, and sense-making.

This emerging fragmented response is not surprising, which brings us to the issue of *scientific uncertainty and controversy*. Teasing out causal chains between the behaviors of financial actors and ecological change are challenging, to say the least. This not only induces heated scientific debates – such as the contested impacts of algorithmic trade on price volatility – but also hampers the emergence of international institutions. Again, based on Young's institutional diagnostics (2008), the fact that the interplay between financial markets, commodity markets, algorithmic trade, price volatility and social–ecological change remain contested and poorly researched effectively undermines any attempt for institutional development. Hence decision-makers at multiple levels focus their energies on polycentric coordination to address urgent and important symptoms (for example, position limits, coordinated action by UN agencies to urgent food crises) rather than on contested and complex underlying drivers (for example, increased demand in world commodity markets due to demographical change and economic development; increases in oil prices).

Politics obviously also plays an important role here as well. Recent controversies in the United States about financial regulation for example, seem to have strong connections to underlying ideological views on the proper role of governments in intervening in the economy and financial markets (Kroszner 2000).

There is another complicating factor here related to the *fragmentation of science*. The lack of collaboration and integrated analysis between financial scholars, global change research and studies of social–ecological systems, is pertinent. Until now, few if any studies have focused on what the increased role of financial actors and hyper-speed trading technologies implies for the production of commodities and associated ecosystem services. The gap between financial experts and ecologists might seem wide, but several recent attempts indicate increased attempts to bridge the breach (Dalton 2005, May et al. 2008, Mandel et al. 2010).

This case also illustrates the main theme of the book: emerging governance challenges at the interface between global environmental, and technological change. All the features of complex change are clearly present: complex systems in terms of adaptive agents in both social (for example, financial) and ecological systems (for example, farmers); connectivity across systems through trade, information and biophysical connections and the associated possibilities of cascades; the potential for surprise due to poorly understood connections (for example, financial markets and commodity price volatility shocks); and lastly threshold change in multiple systems (for example, potential regime shifts in commodity producing landscapes due to changes in production).

What's more, this is an evolving system. As global demands on commodities grow; consumption patterns change, commodity markets become increasingly connected to financial ones; policy-makers respond at international and national levels; and trading technologies continue to evolve, social–ecological–financial interconnections co-evolve over time. Getting a better grip of these part-financial, part-social–ecological connections, their embedded opportunities and risks, will prove critical for governance in a teleconnected planet.

7.   Bridging the 'Anthropocene Gap'

Once in a while you run into a talk, a presentation or a discussion that really helps you put a structure to previously loosely connected ideas or disjointed reflections that you've been pondering fruitlessly for a while. Listening carefully to other peoples' stories usually helps me. My colleague Stephan Barthel is an always unpretentious and insightful storyteller. At a morning seminar about mental models and how they shape the future of our planet, Stephan eloquently elaborated the way different perceptions – or discourses – about the future of urban systems, fundamentally drive urban development *today*. Urbanization is a major force of not only economic development, but also environmental change. The way we choose – or not choose – to design our cities have fundamental impacts on the biosphere. Just consider the aggregated ecosystem impacts of land use change, and changes in surface albedo in a world where an area equivalent to South Africa is projected to be converted to urban land by 2030. This entails a doubling of urban population from today's 2.8 to 4.9 billion (Seto et al. 2012). In addition, more than 60 percent of the cities of 2030 are yet to be built (from Secretariat of the Convention on Biological Diversity 2012).

The interesting part of this story, as my colleague noted, is that there is a tangible tension between very different, yet co-existing urban discourses. Bluntly put, whether policy-makers perceive urban development in the next few decades to be likely to create a 'Planet of Slums' (as proposed by Mike Davis), as a means to advance 'Smart Growth' (as suggested by Howard Frumkin), or as a force harboring an immense stream of diversity, creativity and innovation (as suggested by Edward L. Glaeser), clearly matters. It matters because dominant ideas shape today's perceived policy alternatives, legislative debates, knowledge production, and ultimately physical urban infrastructures and morphologies. As these different perceptions of urban futures grab hold of urban development globally, they lock us into certain human–technological–environmental pathways. Our mental models of the future shape the actions of today.

This might sound overly obvious at first glance, yet I believe that this parallels the wider societal challenges posed by the 'Anthropocene Gap'. Our failure to grasp the complexity of the Earth system, and the tight interconnectedness of global environmental and technological change, is not only of academic interest. It also leads us to continuously restore a sociopolitical infrastructure – including institutions, organizations, and governance processes – inapt to the challenges posed by the Anthropocene.

As mentioned at the outset if this book, the 'Anthropocene Gap' has (at least) three dimensions: a *cognitive*, an *analytical*, and a *political*. I explore each below using the cases elaborated in previous chapters.

A COGNITIVE DIMENSION

The role of mental models, cognitive maps, belief systems, and collective meaning making on decision-making has a long history in the study of agency in politics (Benford and Snow 2000, Campbell 2002). Mental models are also gaining an increased interest in the study of natural resource management (Lynam and Brown 2011). I don't intend to engage with this decades-long debate in detail. What I find interesting in this context, however, is how rigid beliefs and mental models tend to be. Robert Gifford, professor of psychology and environmental studies in Canada, lists seven psychological barriers (or 'dragons of inaction') that impede behavioral changes needed to improve climate change mitigation and adaptation. The first 'dragon' is denoted by 'limited cognition' – that is, our general difficulties in interpreting new risks due to, amongst others, lack of knowledge, ideological biases, perceived uncertainty, and optimism bias (Gifford 2011). These barriers are likely at play as we struggle to grasp with the implications of a new geological era.

I will make it easy for myself and build on the straightforward definition of mental models presented by Timothy Lynam and Katrina Brown (2011, p. 1):

> Mental models represent the way in which people understand the world around them; they are the internal representation of the external system. Mental models are the cognitive structure upon which reasoning, decision making, and behavior are based.

The important connection between mental models and goal-oriented action, are *causal beliefs* – perceptions of the causes of change, and about the actions that can lead to a desired outcome (Milkoreit 2013, p. 34ff). A human-dominated planet – ingeniously and unavoidably wrapped up in highly complex social–ecological and technological systems – poses difficult challenges to simple linear causal beliefs.

Paul Robbins and Sarah Moore (2013) illustrate this tension between older mental models of what constitutes a 'natural' system, and 'conservation' on a human-dominated planet. Robbins and Moore explore 'the nature of recent anxiety, discomfort, conflict, and ambivalence experienced by research scientists in fields confronting ecological novelty in a quickly-changing world' (p. 4), and provocatively speak of the emergence of *ecological anxiety disorder* – the proposed paralysis amongst disciplines such as conservation biology and restoration ecology, created by the rapid loss of 'environmental baselines, grounded and normal conditions from which to make objective assessments for advocating interventions in the world' (p. 8). In short, Robbins and Moore's argument is that as the human enterprise expands its activities and influences all aspects of the biosphere, it also becomes increasingly difficult for ecologists to scientifically define what sort of ecosystems to 'restore' or 'conserve', and to what condition. In their own words, scholars experience an 'increasing fear that past scientific claims about the character of ecosystems and their transformation were overly normative, prescriptive, or political in nature' (p. 9). While the term 'anxiety disorder' might be too strong in the context of this book, Robbins and Moore clearly have an important point worth reiterating: the issue at heart is not scientific uncertainties per se, but also how these entangle with differing causal beliefs and perceptions about alternative desirable futures.

Chapter 4, *Epidemics and supernetworks*, is not only about the intermix between social, environmental and technological factors that drive the emergence of novel and re-emerging zoonotic diseases. It also illustrates the difficulties policy-makers, academia, and the public have in perceiving diseases such as avian influenza, dengue hemorrhagic fever, and malaria as partially ecological challenges related to, for example, land use change. As 'supernetworks' expand as loosely coordinated responses to (perceived) epidemic crises and thresholds, they build on mental models and a 'macro-culture' with shallow ecological understandings.

The chapter *Financial markets, robots and ecosystems* (Chapter 6) also exemplifies difficulties and controversies around causal beliefs as environmental, technological and financial change interplay in ways and at scales never experienced before. As the 2008 food crises and associated debates illustrate, scientists as well as policy-makers are struggling to update (and agree upon) causal beliefs suitable enough to capture unfolding events. Environmental and ecological drivers and impacts remain largely unexplored as scholars (mostly financial economists) attempt to unpack limited aspects of this multifaceted issue (say, 'how much of today's trade in market X and commodity Y is high-frequency trade'). Causal beliefs in both these domains are blurry at best, and faulty at worst, creating tangible institutional gaps.

The notion and current debates about *planetary boundaries* (Chapter 3) is another illuminating example of rapidly changing, and at times conflicting mental models and causal beliefs of the pillars of global sustainability. The notion of 'environmental sustainability' with its legacy in the renowned Rio conference in 1992, had a different focus than today's debate about Earth system resilience. While the former built on a perception of the need to decrease the pressure on natural resources, and minimize waste emissions (Daly 1990), later debates instead focus on the need to steer away from complex interacting planetary 'tipping points'. In the first case, the problem is environmental degradation and resource depletion (say, peak oil); in the latter, a human-dominated planet risking pushing Earth's biosphere into an inhabitable state driven by poorly understood human–environmental feedbacks (say, multiple shocks created by the combination of nonlinear runaway climate change and rapid loss of biodiversity on land and in the oceans).

As already noted, the academic debate is intense on these issues, but the 'Anthropocene Gap' flashes into view once we ask: are these two pre- and post-Anthropocene images compatible (as implied by Griggs et al. 2013), or are they instead representations of two fundamentally different mental models with associated causal beliefs, of the pillars of sustainability? The more I think about these issues, the more I lean towards the second position.

Chapter 5 on geoengineering is probably the most illustrative example of why I think this is the case. The human enterprise already today shapes the biosphere in fundamental ways – ranging all the way from climate change to the speed and direction of biological evolution. As a consequence, our current uses of technologies are already today unintentionally modifying key functions in the Earth system, such as its albedo properties. These two facts are pressing scholars, decision-makers, and others to ask: is it really desirable to intentionally adjust Earth's albedo to rapidly 'cool down' the planet as a means to avoid a climate disaster? What would be a responsible approach to financing, overseeing and evaluating the distributional and environmental impacts of geoengineering experiments and other related approaches (such as de-extinction, large-scale protection of coral reef ecosystems, synthetic biology and artificial photosynthesis) as the renowned 'precautionary principle' fails to provide us with useful guidance? And how do we, as a global society, even begin to grasp two possible alternative futures: a world moving towards a highly risky +3°C to +4°C increase in global mean temperature, and one where some of its worst impacts are mitigated through risky technological interventions?

In my mind, these challenging and contested questions span considerably beyond pre-Anthropocene notions of environmental sustainability.

And the mental models and causal beliefs that we have developed over decades and which have been appropriate before, could now instead put constraints requiring us 'to unlearn previous processes before new ones can be created' (Moynihan 2008, p. 353).

AN ANALYTICAL DIMENSION

Analytical frameworks are critical tools in our attempts to produce simple enough and testable theories of relationships in the world. Emerging debates about the Anthropocene are already provoking discussions in scientific disciplines such as law, ecology and global change science, about the need to rethink and reconfigure existing analytical frameworks (for example, Young et al. 2006, Vidas 2011, Ellis 2011, Folke et al. 2011, Robbins and Moore 2013). As a few scholars already have implied (Young 2012, Biermann 2012), time seems to be more than suitable for a similar debate in political science as well.

By that I mean that some of the standard frameworks that are currently employed to investigate and discuss global environmental change challenges, are falling seriously short. They are failing to provide not only a new generation of political scientists, but also decision-makers and the general public, with much needed insights into the future of environmental politics.

Doubtlessly large social science research initiatives such as the Earth System Governance Project have made important analytical advances in the last few years (Biermann et al. 2012). This progress is particularly clear in research areas such as institutional fragmentation, segmentation and interactions; the changing influence of non-state actors in international environmental governance; novel institutional mechanisms such as norm-setting and implementation; and changing power dynamics in complex actor settings (Young et al. 1998; Biermann and Pattberg 2012; Oberthür and Stokke 2011). The mechanisms which allow institutions and state and non-state actors to adapt to changing circumstances (by some denoted as *adaptiveness*) is also gaining increased – but still modest – interest by scholars (Biermann 2007, p. 333, Young 2010).

Despite this increased interest, very few empirical studies exist that explicitly explore the capacity of *international* actors, institutions and global networks to deal with the sort of complex systems dynamics in human–environmental systems explored in this book. Let me elaborate on this point.

A recent synthesis of critical global environmental governance challenges in the Anthropocene identifies seven important 'building blocks' for institutional reform. These include proposals such as an upgraded UN environmental agency; an improved integration of environmental issues into international policy-making; and closing 'regulatory' gaps related to emerging technologies (just to mention a few) (Biermann et al. 2012). Norichika Kanie and colleagues (2012) argue that a 'transformation of the international institutional architecture for sustainable development' towards 'planetary stewardship' is needed, and present the 'The Hakone Vision on Governance for Sustainability in the 21st Century'. The 'vision' includes a call for broadening of meaningful and accountable participation; improving integration between global-level institutions; and by developing a Sustainable Development Council.

Interestingly enough, both these prominent syntheses that summarize decades of work by social scientists in the field, fail to elaborate the governance mechanisms critical for responding to nonlinear environmental change at global scales. Issues related to the politically contested nature of global threshold behavior; the role of nested multi-level networks and polycentric coordination; and the role of flexibility and experimentation in a new global setting subject to rapid environmental and technological change, are clearly absent. While the complexity lingo is certainly there and can be found in other similar synthesis publications, a deep understanding of its implications has yet to breach into our analytical frameworks.

A First Rough Attempt

What would a renewed analytical framework look like if we strive to contribute to a renewed institutional understanding of the Anthropocene? The short answer is a framework based on polycentricity, and a deeper understanding of the interplay between networks, institutions and Earth system complexity (including human–environmental–technological interconnections).

As several of the cases in this book elaborate (especially the chapters on planetary boundaries and epidemics) the features of global environmental governance required to address, for example, 'tipping point' changes and surprise, are likely to be very different from those needed to tackle incremental (linear) environmental stresses (Folke et al. 2005, Duit and Galaz 2008). The cases analysed in this book not only contain different aspects of complexity – that is surprise elements, possible threshold behavior and cascading dynamics (see Table 7.1). While all are characterized by severe institutional fragmentation, they all also contain different degrees of polycentric coordination, and thus patterns of self-organization in governance.

To be more precise, the issues created by interlinked financial markets (including technologies) and food commodities has until now led to

Table 7.1   Overview of book themes and case studies

Case	Surprise Element	Threshold Dynamic	Cascading Dynamic	Governance Challenges
CHAPTER 3 Planetary Boundaries (PB)	Unexpected shifts in 'boundaries' possible due to changes in scientific understandings and complex bio-geophysical and social interconnections.	Transgression of boundaries might display 'tipping point' behavior (for example, climate change), but scale and implications still contested.	'Boundaries' are interconnected through bio-geophysical linkages and 'telecoupling'. The transgression of some of these (for example, climate) can affect others (for example, biodiversity, water).	Securing the legitimacy of 'planetary boundaries science'; international collective action around dynamic targets characterized by scientific uncertainties; linking suggested institutional reforms to multiple and interacting global environmental stresses.
CHAPTER 4 Emerging Infectious Diseases	Novel infectious diseases with unknown mortality and morbidity; precise location of outbreak hard to predict in advance.	Epidemic thresholds require prompt responses to be able to control spread of the disease.	Epidemic outbreaks can have severe health implications, as well as affect the tourism industry, human security and national interests.	Information processing challenges (securing reliable early warning); building capacity for prompt response across levels of social organization; addressing 'drivers' (for example, land use change, degraded health infrastructure) as well as 'symptoms' (epidemic outbreaks).

Table 7.1 (continued)

Case	Surprise Element	Threshold Dynamic	Cascading Dynamic	Governance Challenges
CHAPTER 5 Geo-engineering Technologies (GET)	Possible surprises in bio-geophysical and ecological responses to deployment of GET; possible unexpected social impacts and responses (for example, public opposition, socio-political tensions).	Critical for climate change in general, but also for related 'tipping elements' such as the Artic sea ice. These affect current discussions on the need to develop and deploy GET.	GET primarily addresses climate change, but also affect other Earth system processes. Deployments of, for example, stratospheric aerosols are likely to affect global precipitation patterns, and indirectly ecosystem processes and functions.	How to create multilevel governance framework/s able to weed out high-risk experiments, but at the same time secure fail-safe innovation of benefit from a PB perspective. Extremely difficult policy trade-offs characterized by very high scientific uncertainty and disagreement, and socio-political tensions.
CHAPTER 6 Algorithmic Trade	Novel and rapid disturbances in financial markets which propagate in to commodity markets such as sugar and cotton.	Presently unknown, but price volatility expected to increase over time due to increased demand and connectedness between international markets.	Presently unknown what the cascading impacts are for human food security and ecosystems, however some anecdotal evidence.	How to effectively manage a rapidly evolving technology with very limited knowledge about impacts. Governance options able to tie financial markets with environmental concerns are non-existent.

fragmented national and international attempts. It has failed to make sense of human–environmental–technological system linkages, and initiate policies able to address contested drivers (say, speculation by large investors with commodity futures). Regulation attempts are 'bottom-up' in the sense that they are driven by individual nations (for example, US legislation and European attempts to increase transparency) or by central actors themselves (for example, self-regulation by large investment banks) with weak (if any) international steering. Clear patterns of stronger polycentric coordination at the international level (explored in Chapter 3), are evidently absent which results in serious mismatches between the scale, speed and scope of the issue, and existing governance attempts.

Geoengineering technologies are also embedded in a fragmented governance setting, but with considerably clearer patterns of polycentric coordination. Emerging international regulation though international bodies such as the Convention on Biological Diversity and the London Convention and Protocol, take place in parallel to emerging attempts for self-regulation.

A similar pattern can be conceived in the case of planetary boundaries. Planetary boundaries and their interactions are notoriously difficult (if not impossible) to pin down through a simple top-down governance approach such as an overarching global agreement. Yet some loose international coordination by international organizations and non-governmental actors including epistemic communities do exist. In addition, some of the core underlying ideas about the risks entailed with nonlinear environmental change seem to be entering arenas of multilateral policy dialogues. The case of epidemics shows even stronger features of strong polycentricity. The International Health Regulations provides a strong 'backbone' for global collective action, but its effectiveness is fundamentally dependent on the operation of 'supernetworks' – and their capacities for information processing, collective intelligence and local to global responses.

These different examples and forms of polycentric coordination might seem disparate, but have something important in common: an intriguing and poorly explored triad between global networks, international institutions, and the perceived possibility of large-scale (regional up to global) surprise and 'tipping points' with probable severe social and ecological cascading consequences.

By shaping state and non-state action, international institutions play a critical role in affecting the creation of potential global change 'tipping points' ((a) in Figure 7.1) (Young 2008, 2010). For example, existing international agreements on biodiversity embedded in the Convention on Biological Diversity influence the way nations try to address biodiversity loss due to deforestation and land use change. As nation states and

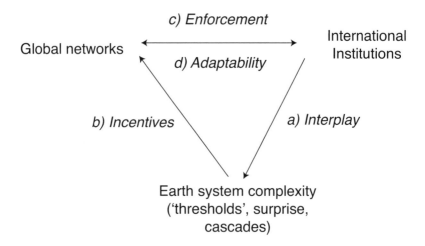

*Figure 7.1   The interplay between global networks, international
institutions and Earth system complexity*

non-state actors meet (or more often, fail to meet) global targets, they
also affect the possibilities for threshold changes in small and large-scale
biophysical systems – or ecosystems.

Normally this is where a conventional institutional analysis would end.
But clearly this is only part of the story. These perceived 'tipping points'
also creates mixed *incentives* ((b) in Figure 7.1) for collective action. While
coordination failure is likely due to actor, institutional and biophysical
complexity (cf. Young 2008 and Chapter 3), the perceived urgency of
the issue can also create *incentives* for action amongst international state
and non-state actors, and spur the emergence of global networks ((b) in
Figure 7.1). These networks can complement or support the *enforcement*
((c) in Figure 7.1) of existing international institutions through their ability
to process information and coordinate multi-network and multi-level
collaboration, as well as create the endogenous and exogenous pressure
needed to induce changes in international institutions. As Young proposes,
these sorts of self-generating mechanisms can help build *adaptability* ((d)
Figure 7.1) and combat 'institutional arthritis' (Young 2010, p. 382). As
scholars such as Olsson and colleagues (2006), and Moore and Westley
(2011) have explored, networks of actors embed an often-ignored potential for large system change by transmitting information, ideas, norms,
or practices. 'System change' includes fundamental institutional transformations, such as shifts to new modes of governance at not only regional
(Olsson et al. 2008), but also global scales (Österblom and Sumaila 2011).

In addition, actors in networks fulfill different functions under changing circumstances, making network dynamics as the environmental and social context changes, a critical issue (Sandström and Ylinenpää 2012).

For example, the emergence of novel zoonotic diseases (such as avian influenza) is intrinsically affected by existing international institutions as these affect the development of urbanization, land use change and technological development (*interplay*). The potential of these diseases to rapidly transgress dangerous epidemic thresholds creates *incentives* for joint action, in this case through the emergence of 'supernetworks', despite malfunctioning formal institutions (for example, International Health Regulations before the year 2005). As nation states agreed to reform the International Health Regulations in 2005, the revisions built on technical standards, organizational operation procedures and norms developed by these loosely WHO-coordinated networks years in advance (*adaptability*) (Heymann 2006).

International actors trying to prepare for the possibly harmful human well-being implications of ocean acidification and rapid loss of marine biodiversity, also illustrate this triad. As these actors perceive the possible transgression of human–environmental 'tipping points' (thereby creating *incentives*), they coordinate their actions in global networks to increase their opportunities to bring additional issues to existing policy arenas created by international institutions (*adaptability*). At the same time, these institutions fundamentally affect the biophysical, technological and social drivers that affect the 'tipping points' at hand (*interplay*, for example, the Convention on Biological Diversity, climate change agreements under the UNFCCC, and the United Nations Convention on the Law of the Sea).

Geoengineering governance initiatives display a similar pattern: as social actors perceive a threat to transgress critical climate and biophysical thresholds (for example, climate and the melting of Artic sea ice), they try not only to develop new technologies, but also build global network alliances and communication networks as they lobby for institutional changes (such as changes in the London Protocol or the Convention on Biological Diversity) to better incorporate the perceived need for experimentation and (in very rare cases) full deployment. Geoengineering opponents mobilize in similar ways, and target the same international arenas thereby creating heated debates in the interface between complex environmental change, technology and politics.

Automatic trade with commodities also displays some loose international coordination driven by concerns of financial surprises and damaging price volatility, and its cascading consequences on, for example, food security. Debates about causality and impacts are heated and polarized, and has until now only led to very weak forms of polycentric coordination.

A Cautious Note

I would like to stress that my suggested focus on this triad is not intended as a functionalist argument. That is, my intention is not to explain the emergence of global networks and international institutions as the mere result of their functions for those who created them (see Pierson 2000b for a good critique). Nor am I arguing that these self-organized global network responses to (perceived) increased Earth system complexity will prove to effectively bridge existing institutional gaps, and remedy 'institutional arthritis'. On the contrary, and as I hope the cases in this book have illustrated, this process is as much about cooperation and coalition building, as about controversies and conflict.

My point is that an attention on this 'triad' puts the complexity of global environmental change at the center stage of analysis. It brings to light the political processes beyond formal multilateral negotiations, and international institutions by highlighting the role of global networks. It brings our much-needed attention to processes of global communication, self-organization, and adaptability at the international level. And lastly, it potentially brings multiple approaches in the social sciences together – such as institutional analysis (Biermann et al. 2012), network analysis (Kim 2013), agent-based modeling (Vasconcelos et al. 2013), studies of policy narratives (Leach et al. 2010), and mental model mapping (Milkoreit 2013) – in ways that contributes to emerging discussions on governance and politics in the Anthropocene.

Governance, Complexity and Technology

What is the role of technological change in all this? As I mentioned in the outset technology and technological change play a central role in this book's endeavor to unpack the 'Anthropocene Gap'.

The first obvious connection is of course that technologies starting with the domestication of tools – for example, fire, plow, irrigation – have profoundly shaped ecosystems and entire regions during the millennia. In a sense, technology is the main vehicle and creator of the 'great acceleration' that eventually brought the human enterprise into the Anthropocene (Steffen et al. 2011, Allenby 2008). Technology also plays a central role in environmental political debates, often discernible in the clash between 'technophilic' and 'technophobic' camps (Brand and Fischer 2012). But these two standard perceptions of technology will clearly not suffice, as this book has intended to demonstrate. Four additional features of technological change need to be considered if we are to start bridging the 'Anthropocene Gap'.

First, perceptions about the impacts of technological change clearly interplay with current controversies about the risks entailed with 'tipping point' changes in the biosphere including the Earth system. As the Chapters 1 Planetary *terra incognita* and 3 *Earth system complexity* elaborated, technological change is on the one hand perceived as a fundamental tool to steer away from major environmental thresholds and disruptions. According to this view, 'planetary boundaries' or Anthropocene doom-and-gloom scenarios are likely to provide ill-founded, and even counterproductive, scientific advice to decision-makers. The opposing position instead states that technological change has not only brought civilization to a situation where dangerous and large-scale environmental threshold changes are imminent, but also could rapidly propagate and transgress geographical boundaries due to increased connectivity created by information and transportation networks. Similar dualistic notions of technological change and thresholds were explored in the chapter *Engineering the planet*: the possibilities for dangerous threshold changes (such as run-away climate change or nonlinear melting of the Arctic) are used as an argument in support of geoengineering experimentation and deployment. On the other hand, the possibility for threshold changes and surprises can also be used as an argument for precaution – that is, intentionally modifying the climate could drive ecosystems beyond dangerous thresholds which forces us to think at least twice, about tinkering with complex systems.

Second, technological change and especially converging emerging technologies pose new governance challenges tightly connected to the Anthropocene debate. Not only has the evolution of communication and transportation networks driven the emergence of 'nested and teleconnected vulnerabilities' or 'cascading ecological crises' (see Chapter 2 *Governance and complexity*). Technologies also co-evolve and converge in ways that create even more profound governance challenges than normally acknowledged. The rapid convergence between financial and commodity markets is indispensably dependent on the evolution of electronic trade, and its speed hinges fundamentally on the development of increasingly advanced algorithms. Current discussions about climate geoengineering would be unmanageable without the existence of highly advanced climate computer models, and global communication networks that bring experts together as they attempt to agree on possible alternative climate futures. The potential 'de-extinction' of species through genetic engineering, the emergence of synthetic biology, novel discussions about the possibilities of artificial photosynthesis, and the start of a possible 3D-printing revolution, are bound to make their imprint in academic and public debates, and spur a different type of environmental governance debate, as we continue to modify the Earth system in fundamental ways.

Third, the distributional impacts of technological change cannot be overseen. Where, how, and in what ways geoengineering technologies (Chapter 5) are developed and (possibly) deployed, have tangible impacts on the distribution of environmental risks and benefits in both space and time. Cooling down Earth through the deployment of stratospheric aerosols might (note: theoretically) benefit a majority of the Earth's inhabitants by reducing global warming, but could redistribute precipitation patterns in ways that put communities directly dependent on rain-fed agriculture and traditional pastoral livelihoods at risk. Geoengineering experiments, such as ocean iron fertilization might seem theoretically sensible as a means of removing carbon dioxide from the atmosphere, but transfers risks to marine ecosystems with potential downstream consequences for other groups. Needless to say, scientific uncertainties are vast here.

These distributional effects can be even less clear. As the chapter on financial markets and ecosystems elaborated (Chapter 6), technological changes in financial markets coupled with the emergence of new players in commodity markets, seems to have created another form of distribution: the distribution of interrelated financial and ecological risk. In short, as trading technologies evolve and financial actors try to reduce their risks by entering commodity markets, new telecouplings push risks to financially smaller actors downstream. If commodity volatilities increase beyond historical analogues, farmers and cooperatives could find themselves making hard choices between long-term investments, or hedging immediate risks. Environmental impacts of these choices are likely, but hard to predict with any confidence. Again, multiple human–environmental causality, scientific uncertainties and controversies, and conflicting causal beliefs make this a hard navigated terrain for scholars and policy-makers alike.

There is also a fourth aspect, and that is the fascinating interplay between information technologies, the expansion of social networks (including organizational networks), and the features of governance. As Jon Pierre and Guy B. Peters note (2005) the governance capacities of states is highly dependent on their abilities for information gathering and processing (p. 46). In this book I've extended this argument, and propose that this also very much applies at the international level. As the chapter *Epidemics and supernetworks* explored, information processing is not only at the very center of multi-network and multi-level collaborations, but also allows centrally placed international organizations to renew their roles and gain legitimacy as global coordinators. As such, they strategically facilitate the creation of 'supernetworks' and collective intelligence, which allow for flexible and prompt responses despite high degrees of strategic and institutional complexity. As I explore in the chapter *Earth system complexity*, communication is also one of the key features of weaker forms of polycentric coordination. The simple expansion of communication networks between international organizations, scientific institutions and additional non-state actors allow for information sharing and coordination, and collective action in the face of complex Earth system stresses.

A POLITICAL DIMENSION

Mark Lynas' book *The God Species: How the Planet Can Survive the Age of Humans* (2011) not only makes for a provocative read, but is probably also one of the few books able to communicate how this (possibly) new geological era is likely to shake up older perceptions about sustainability and 'green' technology. What Lynas fails to investigate is how practically all of the issues touched upon in his book embed fundamental institutional, and therefore also political, issues. Which organizations and through which institutional mechanisms are to define and explore what 'planetary boundary', and at what scale, to include in international environmental policies? Which international organization should be tasked to oversee the development and possible deployment of geoengineering technologies? What legal mechanisms need to be improved at national and international levels to secure the benefits of ecosystem evaluation or water privatization? Becoming intelligent stewards of the biosphere require not only the advancement of technological innovation, but also immense institutional ingenuity. Any examination of Anthropocene challenges that fails to acknowledge its institutional dimensions is deficient at best, and counterproductive at worst.

So far, I have explored the anatomy of the 'Anthropocene Gap' from a cognitive and analytical perspective. Not surprisingly, there is a clear 'political gap' as well. By political here, I mean that current debates about global institutional reforms and policies, have tangible difficulties in grasping the implications of complexity in the Anthropocene era. This leads to (at times) confusing debates where we try to solve tomorrow's problems, with yesterday's solutions. The discussions about needed reforms in the UN-system and the possible development of a new set of global 'Sustainable Development Goals' during and in the aftermaths of Rio+20, are two good illustrations of why we also face a political 'Anthropocene Gap'.

REFORMING THE UN SYSTEM

The need to reform the current United Nations system in ways that more effectively integrates environmental concerns into international policy, organizations and institutions, has been on the international agenda for decades. Several unsuccessful proposals for reform have been put forward by the governments of, for example, Germany, Brazil, Singapore, South Africa, and France over the years (Biermann 2000). The issue gained renewed interest in the run-up to Rio+20 providing what some scholars called a 'constitutional moment' (Biermann et al. 2012) or a 'charter moment, analogous to 1945 when nation states created the United Nations (UN) to deal with issues of peace and security, the most pressing and critical issues at that time' (Kanie et al. 2012, p. 292). Expectations were high, but the supporters of fundamental institutional reform set themselves up for a disappointing result. Rio+20 was consistently plagued by rumors of failure. The deal reached by advance negotiators was critiqued for being too weak to be effective, and Connie Hedegaard, EU climate commissioner, illuminatingly expressed via *Twitter* that 'nobody in that room adopting the text was happy. That's how weak it is' (from Pisano et al. 2012, p. 18).

What I find interesting with these discussions is not whether the outcome of Rio+20 was a relative success or a complete failure. Instead, it is the underlying logic that motivates and drives proposed institutional reforms. As Biermann noted over a decade ago, three major 'deficiencies' often are identified as critical arguments for institutional reform: deficiencies in international coordination between policy arenas, insufficient support for capacity building in developing countries, and inadequate implementation and development of environmental standards (Biermann 2000, p. 24, see also Marshall 2002, Bernstein and Brunnée 2011). These perceived deficiencies in the end set the table for international discussions, and define agreed upon next steps: strengthening the United Nations Environment Programme (UNEP); enhancing the coherence of international support to national sustainable development plans; and securing financial resources and support technology transfer (see Pisano et al. 2012 for details).

Few would question that these are sensible steps in strengthening the effectiveness of global environmental governance. For example, an upgrading and strengthening of the UNEP was the first identified 'building block' towards improved Earth system governance in the Anthropocene identified by Biermann and colleagues (2012) in their authoritative summary.

Clearly, we should also mind the gap created by environmental megaconferences such as these. The reason is that while these proposed institutional reforms might be necessary, they are not likely to be sufficient as we enter the Anthropocene era where connected nonlinear environmental changes and technological innovation reshape the biosphere. This brings (or at least *should* bring) new issues to the table: what new UN-led international knowledge institutions are needed to untangle complex Earth system interactions, and help structure broader societal discussions about the perceived risks of nonlinear environmental change? What international institutional reforms are needed to address not only incremental environmental stresses, but also environmental surprise events that could propagate over national boundaries due to human–environmental teleconnections? What is the role of international actors – such as the UN system and public–private partnerships – as emerging technologies not only make their entrance into heated environmental debates, but also modify social–ecological feedbacks and impact on ecosystem services in ways never experienced before?

Bluntly put: policy-makers need to complement current reform suggestions based on notions such as 'mainstreaming', 'integration', 'capacity-building' and 'participation', with profound discussions about polycentricity, adaptability, innovation and transformation.

CREATING 'SUSTAINABLE DEVELOPMENT GOALS'[48]

Current debates on the possible creation of new international 'sustainable development goals', is another example of similar political aspects of the 'Anthropocene Gap'. The likely future creation of international 'Sustainable Development Goals' (SDGs) is currently high on the international sustainability agenda. While Rio+20 in June 2012 fell short on many issues, the ambition to advance a new set of global sustainability goals is clearly taking root in the UN system through a set of initiatives, such as the High-level Panel on the Post-2015 development agenda, and the intergovernmental Open Working Group on SDGs.

One common view is that the future SDGs should be firmly based on lessons from the development and implementation of the Millennium Development Goals (MDGs) (for example, Sachs 2012). Just as a short recapitulation, the MDGs were established in 2000, and contain eight international development goals to be addressed within a 15-year time frame by the international community. These include eradicating extreme poverty and hunger, reducing child mortality, and ensuring environmental sustainability.

The new SDGs have the ambition of strengthening and integrating the environmental dimension of the previous MDGs. Discussions between national governments and non-governmental organizations are currently intense in different policy arenas on what specific goals and associated indicators to include.[49] The importance of the final choice should not be underestimated, as the goals are likely to draw considerable attention and resources from the international community in the next decade.

However, transferring insights from the MDGs to the SDGs will do little to bridge the 'Anthropocene Gap'. Three issues or dilemmas make this point clear – the integration of 'planetary boundaries' into the SDGs; the role of 'good governance'; and defining environmental goals on a rapidly changing planet.

SDGs, 'Planetary Boundaries' and Thresholds

The risks entailed with nonlinear changes in ecosystems, the Earth system, and 'planetary boundaries', have gained considerable policy attention in the last few years (Galaz et al. 2012b). PBs have also repeatedly been proposed to become a central aspect of the future SDGs (Leach et al. 2012, Sachs 2012, Griggs et al. 2013).

While the transfer of 'planetary boundaries' research into international policy arenas often is implicitly assumed to be a rather straightforward science communication challenge, thresholds of this sort are much more than scientifically informed numerical estimates. As Chapters 1 and 3 elaborated, biophysical thresholds also spark intense political and scientific debates with striking similarities to older political conflicts created by the perceived tension between environmental protection and economic growth and development. The intense international media and political debate about 'planetary boundaries' at the run up, and during Rio+20, is a vivid illustration of the international politics of threshold uncertainty as social interests mobilize around different perceptions and opinions of the dangers of large-scale threshold change.

This creates the first tangible dilemma in the creation of SDGs: while 'planetary boundaries' might be seen as a useful operationalization of the development challenges posed by the Anthropocene, the concept's transition into international policy arenas will be far from painless. While this certainly is no surprise to scholars studying the interplay between scientific advice and international policy, there is a much deeper science–policy dilemma here: is it possible to create global goals able to capture the risks involved with potential very large-scale biophysical threshold changes, but that also are sensitive to likely controversies unleashed by threshold uncertainty? So far, this issue has been outside of the agenda amongst policy-makers, experts and sustainability scientists trying to contribute to the discussions.

SDGs and 'Good Governance'

'Good governance' – simply defined as accountability, government effectiveness, rule of law, and control of corruption (Kaufmann 2003) – is often identified as one key pillar in global attempts to reach the MDGs, and is also promoted as a key strategy for attaining the future Sustainable Development Goals (Sachs 2012, p. 2208). Few would disagree with the view that advancing 'good governance' is a desirable mission. The question is how these indicators and associated policies relate to the sort of human–environmental complexity explored in this book.

Decades of studies of so-called social–ecological systems (Folke et al. 2005, Folke et al. 2011), have stressed that even the most effective, non-corrupt and accountable governments tend to establish institutions and policies which promote efficiency, over diversity and resilience to changing circumstances. More specifically, reforms promoting 'good governance' could utterly fail to acknowledge the key role played by local ecological stewards and knowledge, undermine effective co-management regimes engaging a wide diversity of stakeholders, and implement 'top-down' natural resource management institutions ill-suited for local social and ecological realities (often denoted as 'panaceas', Ostrom 2007). These governance panaceas have historically resulted in the transgression of critical ecological thresholds, often with damaging impacts on human well-being (Holling and Meffe 1996, Folke et al. 2005, Millennium Ecosystem Assessment 2005).

This creates a *second dilemma*: while 'good governance' certainly might be a desirable (and measurable) goal, and a sensible strategy for addressing certain forms of environmental stresses, it is far from a sufficient governance requirement in settings where environmental and ecological change is characterized by continuous change, surprise and nonlinear properties (Folke et al. 2011). Managing ecological phenomena like these requires multi-level institutions and collaboration processes nimble enough to adapt to environmental surprise and change, rather than shallowly efficient top-down blueprint interventions. These latter governance processes – many of which have been explored throughout this book and by colleagues for decades (see summary in Folke et al. 2005) – are comparatively more difficult to capture in simple numerical indicators. Consequentially, they lack an appropriate global data set making them less attractive for the sort of global target setting foreseen for the SDGs. Complementing and exploring the features of 'good governance' in ways that match a new Anthropocene setting is a critical policy challenge.

SDGs and Environmental Goals on a Changing Planet

The Anthropocene is not only characterized by pervasive ecological change, but also by the transformation of ecosystems into new configurations with few and contested historical analogues sometime denoted as 'novel ecologies' (Ellis and Ramakutty 2008, Hobbs et al. 2009). In addition, ecosystems have not only been modified, but are also in transition due to environmental change (for example, climate change) and human induced modifications (for example, expansion of agriculture). Operationalizing a subset of concrete, and tangible global targets that encompass 'Environmental sustainability' will prove extremely challenging in such a dynamic setting, an observation seldom made in current discussions about future possible SDGs. The discussed design of biodiversity targets, and the nonlinear features of ecological change, are illuminating examples of this challenge.

Biodiversity targets: One suggested way to successfully encapsulate 'Environmental sustainability' in the SDGs is to infuse them with environmental targets drawn from existing international agreements, for example the Aichi Biodiversity Targets (Griggs et al. 2013). While these 20 targets have been carefully designed to underpin state action, many of them are illustrative of what are likely to become highly conflictive issues in the near future. For example, while the Aichi targets speak of controlling and removing invasive species (Target 9), the Anthropocene is characterized by an increased presence of 'novel' ('emerging' or 'no-analog') ecosystems that have been generated by a combination of species invasions and environmental change such as climate change (Seastedt et al. 2008, Hobbs et al. 2009). As Norberg and colleagues show (2013), global changes in the diversity and distribution of species are likely to be considerable as species migrate, compete and co-evolve due to climate change.

In many cases, invasive species have no discernible effects to the function of ecosystems, while in other situations they play an important role, by providing habitat and resources for many other species. There is even the further possibility that such non-native species could provide important ecosystem services in the future. This has induced intriguing questions amongst ecologists: what is a 'novel' or 'invasive' species on a planet so fundamentally shaped by human action? And does it make sense to set up environmental goals based on a notion of ecosystems as particular assemblages of species, in particular places, at particular times?

Nonlinear ecological change: The rapid expansion of the extent and intensity of human activities is also increasing the possibility of abrupt, difficult to reverse changes that alter the supply of the ecosystem services upon which societies depend (Biggs et al. 2009). These so-called *regime shifts* can be characterized by fundamental reconfigurations in ecological attributes and flows of ecosystem services that affect society (Millennium Ecosystem Assessment 2005).

Scientific uncertainties are plentiful, and have (as already explored in previous chapters) induced new debates about the global risks of abrupt cascading ecological changes. Conservation scientists, ecologists and managers are also engaged in fundamental debates about the most effective ways to 'protect' the world's ecosystems as the world enters the Anthropocene. For example, classic ecological goals such as restoring and conserving pristine ecosystems are being challenged to change focus from 'damage control', to 'unconventional' and 'non-passive' management. These latter forms of management implies the deployment of large-scale technological interventions, such as the creation of protective shields, artificial reefs and gene-banks to help restore coral reefs ecosystems in the face of ocean acidification and climate change (Rau et al. 2012, see also Nyström et al. 2012).

These emerging debates are likely to sooner or later force policy-makers to not only agree upon and implement sustainability goals, but also to seriously engage in intense debates about the opportunities, risks and distributional consequences of 'non-passive' management of ecosystems. As explored in *Engineering the planet* (Chapter 5), these are highly politically charged issues that could increase in intensity as technological change expands human capacities to modify and engineer atoms, cells, species, ecosystems and even the climate system.

Hence the creation of new international sustainability goals faces serious challenges. While fixed, quantitative (measurable) environmental goals and targets facilitate political action, static environmental targets can easily become obsolete in a rapidly changing environment archetypal for the Anthropocene. In addition, the notion that these targets are most effectively implemented by promoting 'good governance' ignore the need for adaptive modes of governance which connect international and national policies to local social and ecological circumstances, and which also are able to promote flexibility to environmental surprise and change. The cognitive and analytical gaps explored early in this chapter might seem extensive, but the political gap is probably the widest. Where do we even start in our endeavor as political scientists to help bridge this political gap? This brings us back to the 'governance puzzles' presented in Chapter 2 *Governance and Complexity*. As the reader will discover immediately, there are no easy answers, but the chapters have hopefully contributed to a deeper and more informed perception of the issues, and their importance.

EXPLORING THE PUZZLES

Governance Puzzle (1). What characterizes international institutions that are able to detect and respond to 'global human–environmental surprises' of large importance to human well-being?

How do we tackle the fact that some of the consequences of environmental change in the Anthropocene will unfold as 'surprise events' due to complexity and human–environmental connectivity? As several of the chapters have explored, part of the answer can be found in the intricate interplay between loosely coordinated global networks, international institutions and the existence of a shared vision and strategy created by 'macro-culture'. This becomes overly clear in the chapter *Epidemics and supernetworks* that explored the ways in which a few central actors manage to facilitate the actions of multiple, diverse and multi-level networks. These 'networks of networks' or 'supernetworks' are fundamentally kept together by their remarkable capacities for information gathering, processing and dissemination. At the same time, they are embedded and influence formal international institutions as a means to secure resources, gain internal and external legitimacy, and facilitate polycentric coordination.

Governance Puzzle (2). Are international institutions at all able to address complex Earth system interactions, or should we instead put our faith on the emergence of polycentric approaches?

The complex and contested nature of environmental changes pose severe challenges to institutional emergence and design. This is especially a challenge considering the present fragmented nature of global environmental governance. An increasing body of research therefore explores the possibilities entrenched in 'bottom-up', or polycentric approaches. As the chapters *Epidemics and supernetworks* and *Earth system complexity* explored in detail, polycentric approaches entail a number of interesting properties which can help remedy institutional fragmentation and segmentation: they can support communication across sectors and levels; they can help coordinate action which build adaptive capacities; and provide a setting for experimentation and learning. Modeling approaches also show that polycentric initiatives as an aggregate can be more effective than a global agreement. These approaches will however face problems if causal beliefs are contested, issues become conflictive due to diverging interests, or political circumstances change. In addition, dealing with the dynamic nature of global environmental change is likely to require institutional changes related to the role of international environmental assessments, the mandate of international organizations, and managing institutional interactions. Hence polycentric approaches cannot replace, but rather complement international institutions in important ways (see Chapter 3 and concluding analysis in this chapter).

Governance Puzzle (3). Is a governance setting possible that is strong enough to 'weed out' technologies that carry considerable ecological risk, but still allows for novelty, fail-safe experimentation and continuous learning?

Technological change and increasing concerns about the possible implications of threshold changes on ecosystems and the biosphere is inducing new discussions about the role of experimentation, precaution and governance. These debates are inseparably part of the Anthropocene debate, and urge social scientists to have a new look at the political and governance challenges posed by technological innovation and change. Until now, discussions have centered on the features of a proper international framework. Yet if we put governance for complexity, and an interacting Earth system at the center of analysis, other issues arise. These are related to questions about whether 'creative experiments that cover scales and that can fail safely as new possibilities are created and tested' (Holling 2004) can be done in ways that engage with local ecosystem stewards, and promote technologies that not only address climate stresses, but could also bring multiple social–ecological benefits to communities. As mentioned in the chapter *Engineering the planet*, the standard 'precautionary principle' will provide little if any guidance in this context. A governance setting, which allows fail-safe experimentation, and continuous learning, is conceivable in principle. It will require considerable ingenuity, recognition of the socially contested nature of emerging technologies and intentional experimentation with 'natural' systems, and a governance focus on polycentricity.

Epilogue   Back to London via the Baltic Sea

Will we ever be able to 'bridge' the cognitive, analytical and political gaps created by our inability to grapple, analyse and respond to the major implications induced by the Anthropocene? The question haunted me throughout the whole conference in London in 2012. Several presentations and discussions seemed to add to the bigger puzzle. The majority however provided clear illustrations of how even highly skilled sustainability scientists of all sorts, failed to see that the analysis and solutions of today were ill-suited for a new context where technological and human–environmental change is drastically transforming the surface and function of planet Earth.

One optimistic answer would be that it is only a matter of time before we – the public, decision-makers, environmental activists, business leaders and scholars – rise up to the challenge. The Anthropocene debate is merely a few years old, and human ingenuity never ceases to amaze. Needless to say, examples of where science is ignored and misused; where technological innovation has led to detrimental impacts on vulnerable groups in society; or where important powerful interests have blocked preventive decisions, are plentiful (for an interesting discussion, see European Environment Agency 2013). And as Marten Scheffer and colleagues explore (2003, Scheffer and Westley 2007), multiple mechanisms can contribute to delayed societal responses to new environmental problems. One will prove critical for the 'Anthropocene Gap' – problems created by phenomena, linkages or processes unlike those experienced in the past, tend to remain undetected for considerable time due to mental 'lock-in'.

As I was concluding this book, a follower on the micro-blog service *Twitter* made me aware of an interesting and emerging political discussion about the environmental goals for the Baltic Sea. The Baltic Sea is a sea bounded by the Scandinavian Peninsula (Sweden, Finland), the mainland of Europe (in other words, Germany, Poland, Russia and the Baltic States), and the Danish islands. At the beginning of 2013, a well-known Swedish environmental science journalist started to bring the results from different expert reports together, and noted that the multilateral political goal to return to the ecological state of the Baltic Sea in the 1950s would have serious implications for Swedish environmental policies. In short, if the recommendations based on modeling results were followed, Sweden would have to basically shut down all human activities affecting the Baltic Sea drainage basin. But this wouldn't be enough – it would probably also require an active net removal of 140 tons of phosphorus yearly (Jewert 2013a, b).

Human activities around the Baltic Sea drainage basin are far from insignificant, and include agriculture, forestry, and other sources of nutrients stemming from the area's 90 million multinational inhabitants (of which 9 million live in Sweden). Not surprisingly, interest groups such as the Baltic Farmers Forum on Environment now demand a total rethink of existing environmental goals,[50] in their turn drawing intense criticism from environmental NGOs.[51]

The heart of the matter seems to be the combination of scientific uncertainty, thresholds and social conflict. First, drastic reductions in phosphorous leakage could theoretically bring the Baltic back to its pre-industrial ecological conditions. But it would take about 50–100 years (Elmgren 2012, the precise figure is contested, however). Second, the Baltic has been so drastically modified by human activity through overfishing, pollution and eutrophication over the last few decades that the system is likely to have transgressed a bio-geophysical threshold, making a smooth recovery to its previous state practically impossible (Österblom et al. 2007). A few scholars at the fringes of the debate have suggested that such a return therefore also would require unprecedented technological intervention, such as pumping massive volumes of oxygen-saturated surface water to the deep-water to help combat severe anoxic bottoms (Stigebrandt 2012, for an opposite opinion see Conley 2012).

Still, national and multilateral environmental targets, environmental NGO campaigns, and national policy discussions implicitly continue to center around the notion that recovery is possible through carefully negotiated incremental nutrient reductions. In addition, the European Union (through the Water Framework Directive) wants high water quality to be achieved by 2015 in coastal waters, and by 2020 in open sea areas (through the Marine Directive).

Should the idea of a return to pre-industrial Baltic Sea be abandoned as a policy goal? What would a new vision look like in that case, considering its current condition with vast areas of dead bottoms and repeated summer blooms of cyanobacteria? Does the system really contain a hard-wired threshold; can we push the system back; and how long would it take for it to recover? A few years, decades, or hundreds of years? Are technological interventions needed, and at what scale? Should governments around the

Baltic advance larger scale ecological experiments in the region to explore alternative interventions? How would these be funded, regulated and over-seen? What would be the proper political multilateral arena be to advance a Baltic vision adapted to the realities of a human-dominated planet? Or are these debates simply distractions presented in a time when immediate preventive action is needed? These contentious questions provide an addi-tional illustration of the interplay between the Anthropocene, technologi-cal change and politics explored throughout this book.

Readers looking for simple answers, policy guidelines or one dominant framework to guide us through issues like these will surely feel discontent with the sketchy, messy and multifaceted picture drawn in this book. As a complexity scholar, I always have great problems in delivering catchy policy solutions, or blueprints to the social–ecological–technological phe-nomena such as those explored in this book.

Despite this, I remain an optimist. I often find myself agreeing with techno-optimists and their ever-enthusiastic observation that humans have continuously pushed the frontiers of what is perceived as possible. Not because I find their arguments over-compelling (they seldom are), but rather because I'm truly fascinated by the enormous creativity I'm fortunate to encounter almost daily in enthusiastic scholars and students, energetic entrepreneurs, intelligent artists, ambitious teachers and engaged activists of all sorts. Surely uncertainties are vast. One thing I know for sure – time is ripe for a very different discussion.

Notes

1. This line of argument has been questioned recently by Lenton himself (Lenton 2013).
2. Two short, but very important definitions are worth bringing out here. By 'governance' I mean the multi-level patterns of interaction among actors, their sometimes conflicting objectives, and instruments besides institutions that are chosen to steer social and environmental processes within a particular policy area (cf. Pierre and Peters 2005). By 'institutions' I mean 'shared concepts used by humans in repetitive situations ordered by rules, norms and strategies' (Ostrom 2007, p. 23). The emphasis in this book will be on formal rules (for example, national laws, international regulation and similar), rather than norms. Governance in this book hence entails a wider view on societal steering than institutions, but is also fundamentally affected by such. Note also that 'institutions' and 'organizations' describe distinct social phenomena.
3. This term is very much inspired by Thomas Homer-Dixon's excellent book *The Ingenuity Gap – Facing the Economic, Environmental and other Challenges of an Increasingly Complex and Unpredictable World*. New York: Vintage Books (2002).
4. I use the same definition of 'Earth system' as defined by Steffen et al. (2007, p. 615), in other words, 'the suite of interacting physical, chemical and biological global-scale cycles and energy fluxes that provide the life-support system for life at the surface of the planet'. Throughout this book, I will also use the term 'biosphere' as the sum of all the planet's ecosystems (Lenton and Williams 2013, p. 381).
5. *New Scientist*, 'Wind and wave farms could affect Earth's energy balance', 30 March 2011. Note that this calculation is intensively debated by the scientific community, for example, M.Z. Jacobson and C.L. Archer 'Comment on "Estimating maximum global land surface wind power extractability and associated climatic consequences"'. Available at http://www.earth-syst-dynam-discuss.net/1/C84/2010/esdd-1-C84-2010-supplement.pdf (accessed 8 January 2013).
6. This creates difficult trade-offs and novel interaction between governance and emerging technologies that I will explore in Chapter 5 about geoengineering.
7. Which of these are truly nonlinear and entail threshold behavior, and at what scale, is debated (see below and Chapter 3). In this book (and probably to the frustration of many colleagues), I will use a very simple definition: by 'threshold', 'tipping point', and 'nonlinear change', I mean an abrupt breakpoint between alternate states of a system, where a small change or disturbance produces a large change in the characteristic structure, function and feedbacks of the system.
8. In 'Planetary boundaries as a power grab', 4 April 2013, online: http://rogerpielkejr.blogspot.se/2013/04/planetary-boundries-as-power-grab.html?spref=tw.
9. In the comments field of 'Planetary boundaries as a power grab', 4 April 2013, online: http://rogerpielkejr.blogspot.se/2013/04/planetary-boundries-as-power-grab.html?spref=tw (accessed 10 September 2013).
10. See Galaz, V. 'A planetary boundaries straw-man', 8 April 2013, online: http://rs.resalliance.org/2013/04/08/a-planetary-boundaries-straw-man/ (accessed 10 September 2013).
11. 'The re-emergence of the neo-Malthusians'. Available at http://rogerpielkejr.blogspot.se/2012/04/reemergence-of-neo-malthusians-guest.html (accessed 9 November 2012).
12. Defining 'technology' is an immense task in itself of course. In this book I will use the threefold definition presented by Brian Arthur (2009, pp. 28–29): 'technology' is a means to fulfill human experience; as an assemblage of practices and components; and

as the collection of devices and engineering practices available in a culture. The definition is broad, but captures all aspects of technology explored in this book.

13. This section builds on a synthesis presented in Duit and Galaz (2008).

14. It should be noted that not all biophysical complex systems are adaptive. For example, once the Greenland ice sheet crosses a threshold where melting is losing more ice than precipitation is delivering, it will eventually disappear. It can't 'adapt'. Thanks to Will Steffen for pointing this out.

15. There have also been a number of parallel attempts in the social sciences to analyse the nonlinear nature of social and political behavior (for example, Jervis 1997, Pierson 2000a).

16. For space reasons, I will explore these phenomena only briefly. For further reading I recommend Homer-Dixon (2002) and Norberg and Cumming (2008).

17. As a reminder, I use a very simple definition of 'thresholds' – an abrupt breakpoint between alternate states of a system, where a small change or disturbance produces a large change in the characteristic structure, function and feedbacks of the system.

18. The '2 degree target' refers to a number of political decisions (for example, at the UNFCCC conferences of parties (COP) in Copenhagen (COP-15) and Cancun (COP-16)) to adopt a climate stabilization target with the aim of limiting global warming to 2°C above the pre-industrial level. This target is sometimes, but not always, framed as a threshold limit within which social adaptation within 'acceptable' economic, social and environmental costs is possible. See Knopf et al. (2012) and the EU Climate Change Expert Group (2008) for details.

19. Anne-Sophie Crépin, Juan Carlos Rocha, Maja Schlüter and Emilie Lindkvist (all associated to the Stockholm Resilience Centre), have provided important initial feedback to this section.

20. It should be noted that the notion of 'adaptive governance' has its roots in several subfields, including studies of adaptive management (Holling 1973), adaptive co-management (Berkes et al. 1998, Plummer et al. 2012), analysis of CPR-institutions (Ostrom 2005), and polycentric governance (Ostrom 2005). In addition, the field is clearly inspired by concepts, advances and methodologies developed in associated research streams such as social network theory, policy studies, innovation research and organizational theory (see Folke et al. 2005 and Olsson et al. 2006 for examples).

21. It should be noted that an analysis still needs to acknowledge the role of institutions and social behavior at not only global, but multiple levels of social organization. International institutions and global governance consistently influence and are influenced by lower level governance processes (Hooghe and Marks 2003), an observation that is acknowledged throughout this book.

22. The concept is contested: Dimitrov also uses the international forest negotiations as a case due to the lack of binding formal international treaty, while others point out that there indeed exists a forest regime when looking at the broader forest governance setting. Thanks to Gunilla Reischl for pointing this out.

23. See Lacey (2008) for an awakening reminder.

24. This section builds on the results of a workshop presented in Galaz, Biermann, Crona, Loorbach, Folke and Olsson (2012). I would like to acknowledge the critical contribution made by colleagues such as Frank Biermann, Måns Nilsson, Per Olsson and Karin Bäckstrand to the thoughts summarized in this section.

25. My emphasis here is on polycentric *order*, rather than *systems*. As the next sections will explore, I believe that the term order is analytically more useful to describe the suite of coordination and collaboration mechanisms that are at play at the international level.

26. 'H' and 'N' are used to classify different types of influenza A. The H stands for haemagglutinin and the N stands for neuraminidase. Both are proteins on the surface of the virus, each of which is given a different number.

27. This chapter builds on already published work (Galaz 2009, Galaz 2011) with some important additions from unpublished interviews with key coordinators of epidemic warning systems, as well as ongoing work in the international research project 'Dynamic Drivers of Disease in Africa' (http://steps-centre.org/project/drivers_of_disease/).

28. These 'thresholds' are hard to quantify, are usually identified as the point where the spread of disease becomes increasingly difficult to control if the so-called basic reproduction number R0 >1. This value is defined by for example, virulence, transmissibility and severity: see Wallinga and Teunis (2004). Another way to understand these are the WHO pandemic phases, with each step in the ladder implying a qualitative shift in disease risks. See WHO (2005).

29. The Hindu Universe, 25 September 1994, 'Plague hits hard in Western India'. Available at http://www.hindunet.org/alt_hindu/1994/msg00686.html (accessed 18 September 2013).

30. This section builds on interviews conducted in 2009, with Daniel Beltran Alcrudo at the Global Early Warning System for Major Animal Diseases (GLEWS) at FAO in Rome (February 2010), and Michael Blench, Technical Advisor and Project Coordinator, GPHIN at Health Canada in Ottawa (March 2010).

31. Based on an interview with Michael Blench at GPHIN, see previous endnote.

32. AFP 29 April 2009 'Debate rages over swine flu name' available at http://www.google.com/hostednews/afp/article/ALeqM5g0E8OKGtLJZKNrYKzA1ilBdLV6yQ.

33. *The New York Times*, 28 April 'The naming of swine flu, a curious matter' available at http://www.nytimes.com/2009/04/29/world/asia/29swine.html.

34. This chapter builds on Galaz (2012), and my first-hand experiences as one of the lead authors for the Convention on Biological Diversity's Expert Liaison Group on Geoengineering. I would like to thank the members of this expert group for their insights during our repeated discussions in London, and especially Andrew Parker (Harvard University) and Ralph Bodle (Ecologic Institute) for their insights. The last parts of this chapter have benefited considerably from repeated discussions with Jason Blackstock (University of Oxford). Any potential flaws in the argument are exclusively my own.

35. Readers familiar with the geoengineering debate are recommended to skim through the first two parts, and focus on Part 3.

36. Again, this is a very rough simplification. For example, carbon dioxide removal technologies such as ocean iron fertilization could also be portrayed as containing high risks. See Secretariat of Convention on Biological Diversity (2012), Royal Society (2009) for details.

37. Available at http://www.iisd.ca/download/pdf/enb09544e.pdf (accessed 13 September 2013).

38. This section builds on Cressey (n.d.), Specter (2012), and Black (2011).

39. From *The Engineer*, 'Particle injection could abate climate change', by Siobhan Wagner 16 July 2010. Available at http://www.theengineer.co.uk/news/news-analysis/particle-injection-could-abate-climate-change/1003647.article (accessed 14 September 2013).

40. This section builds on the following news articles:
'World's biggest geoengineering experiment "violates" UN rules', by Martin Lukacs, *Guardian*, 15 October 2012. Available at http://www.guardian.co.uk/environment/2012/oct/15/pacific-iron-fertilisation-geoengineering (accessed 17 June 2013).
'Its livelihood gone, fishing village tries scheme to seed a barren sea', by Mark Hume, *The Globe and Mail*, 16 October 2012. Available at http://www.theglobeandmail.com/news/british-columbia/its-livelihood-gone-fishing-village-tries-scheme-to-seed-a-barren-sea/article4617262/ (accessed 17 June 2013).
'Replacing iron not pollution, village says', by Judith Lavoie, *Times Colonist*, 20 October 2012. Available at http://www.timescolonist.com/health/Replacing+iron+pollution+village+says/7421187/story.html (accessed 17 June 2013).
'Ocean fertilization could be a boon to fish stocks', by Ron Johnson, *Earth Island Journal*, 31 October 2012. Available at http://www.earthisland.org/journal/index.php/elist/eListRead/ocean_fertilization_could_be_a_boon_to_fish_stocks (accessed 17 June 2013).
'Dumping of iron into sea off Haida Gwaii suspended amid acrimony', by Judith Lavoie, *Times Colonist*, 23 May 2013. Available at http://www.timescolonist.com/news/local/dumping-of-iron-into-sea-off-haida-gwaii-suspended-amid-acrimony-1.229839 (accessed 17 June 2013).

41. Quote from 'Replacing iron not pollution, village says', by Judith Lavoie, *Times Colonist*, 20 October 2012. Available at http://www.timescolonist.com/health/ Replacing+iron+pollution+village+says/7421187/story.html (accessed 17 June 2013).
42. See for example, Vidal (2012), and Kintisch (2010).
43. Quote from Vidal (2012).
44. See for example ETC Group (2009).
    Greenpeace, 'Geoengineering the ocean off Haida Gwaii: A false solution to real problems' available at http://www.greenpeace.org/canada/en/high/Blog/geo-engineering-the-ocean-off-haida-gwaii-a-f/blog/42855/ and 'Open Letter to the IPCC' signed by 167 organizations, 2011. Available at http://www.handsoffmotherearth.org/wp-content/ uploads/2011/08/IPCC_Letter_with_Signatories_-_7-29-2011.pdf.
45. The work presented here has been developed in collaboration with my colleagues Fredrik Moberg (Stockholm Resilience Centre), Johan Gars (Beijer Institure), Björn Nykvist (Stockholm Resilience Centre), and Jakob Lundberg (previously at Food and Agricultural Organization Nordic).
46. It should be noted that the definition of AT in itself, is contested. Here I define algorithmic trade (AT) as trade that relies on high-speed computer linkages; uses of sophisticated statistical, econometric, machine learning, and other quantitative techniques; and cases where traders hold positions for very short periods of time (micro-seconds) from GOS (2011, p. 28), see also US Securities and Exchange Commission (2010), Hendershott et al. (2011, p. 1).
47. I use the same standard definition of 'price volatility' as that used in the economic literaure, that is 'the standard deviation of price returns, in other words, the standard deviation of changes in logarithmic prices', from Gilbert and Morgan (2010, p. 3024).
48. This section is based on discussions and a co-authored paper with my colleagues Albert Norström (Stockholm Resilience Centre) and Martin Sjöstedt (Göteborg University).
49. UNSD Secretariat, Rio+20 Issues Briefs, 'Current ideas on sustainable development goals and indicators', No 6. Available at http://www.uncsd2012.org/content/docum ents/218Issues%20Brief%206%20-%20SDGs%20and%20Indicators_Final%20Final% 20clean.pdf (accessed 19 September 2013).
50. See letter 'Viewpoints from the farmer organisations around the Baltic Sea to the proposal for the ministerial declaration concerning revised HELCOM Baltic Sea Action Plan (BSAP)', 17 June 2013. Available at http://www.lrf.se/PageFiles/422/ BrevBFFE%20detailed%20viewpoints%20to%20HELCOM%20dated%2017%20June %202013.pdf (accessed 3 September 2013).
51. Letter from Lennart Gladh at the WWF 'Synpunkter på LRFs förslag till ny Östersjöpolitk' (undated) available at https://docs.google.com/file/d/0B0H9KDhnWEq-bGFiYXNnSlZzQ3M/edit?usp=sharing (accessed 5 September 2013).

References

Adger, W.N., H. Eakin and A. Winkels (2009), 'Nested and teleconnected vulnerabilities to environmental change', *Frontiers in Ecology and the Environment*, **7** (3), 150–57.

Aligica, P.D. and V. Tarko (2011), 'Polycentricity: from Polanyi to Ostrom, and beyond', *Governance*, **25** (2), 237–62.

Allen, M. (2009), 'Tangible boundaries are critical', *Nature Climate Change*, **3** (0910), 114–15.

Allenby, B. (2008), 'The Anthropocene as media: Information systems and the creation of the human Earth', *American Behavioral Scientist*, **52** (1), 107–40.

Allendorf, F.W. and J.J. Hard (2009), 'Human-induced evolution caused by unnatural selection through harvest of wild animals', *Proceedings of the National Academy of Sciences of the United States of America*, **106** (Suppl 1), 9987–94.

Alley, R.B., J. Marotzke, W.D. Nordhaus, J.T. Overpeck, D.M. Peteet, R.A. Pielke Jr., R.T. Pierrehumbert, P.B. Rhines, T.F. Stocker, L.D. Talley and J.M. Wallace (2009), 'Abrupt climate change', *Science*, **299** (5615), 2005–10.

Allison, H.E. and R.J. Hobbs (2004), 'Resilience, adaptive capacity, and the "lock-in trap" of the Western Australian agricultural region', *Ecology and Society*, **9** (1), 3.

Anderies, J.M., P. Ryan and B.H. Walker (2006), 'Loss of resilience, crisis and institutional change: Lessons from an intensive agricultural system in southeastern Australia', *Ecosystems*, **9** (6), 865–78.

Andonova, L.B., M.M. Betsill and H. Bulkeley (2009), 'Transnational climate governance', *Global Environmental Politics*, **9** (2), 52–73.

Andreae, M.O., C.D. Jones and P.M. Cox (2005), 'Strong present-day aerosol cooling implies a hot future', *Nature*, **435** (7046), 1187–90.

Angel, R. (2006), 'Feasibility of cooling the Earth with a cloud of small spacecraft near the inner Lagrange point (L1)', *Proceedings of the National Academy of Sciences of the United States of America*, **103** (46), 17184–9.

Ansell, C. (2006), 'Network institutionalism', in Binder, S., R. Rhodes and B. Rockman, *Oxford Handbook of Political Institutions*, Oxford: Oxford University Press.

Arsel, M. and B. Büscher (2012), 'NatureTM Inc.: Changes and continuities in neoliberal conservation and market-based environmental policy', *Development and Change*, **43** (1), 53–78.

Arthur, W.B. (2009), *The Nature of Technology: What it is and how it evolves*, New York: Free Press.

Axtell, R.L., J.M. Epstein, J.S. Dean, G.J. Gumerman, A.C. Swedlund, J. Harburger, S. Chakravarty, R. Hammond, J. Parker and M. Parker (2002), 'Population growth and collapse in a multiagent model of the Kayenta Anasazi in Long House Valley', *Proceedings of the National Academy of Sciences of the United States of America*, **99** (1), 7275–9.

Banerjee, B. (2011), 'The limitations of geoengineering governance in a world of uncertainty', *Stanford Journal of Law, Science and Policy*, **240** (May), 15–36.

Barnosky, A.D., E.A. Hadly, J. Bascompte, E.L. Berlow, J.H. Brown, M. Fortelius, W.M. Getz, J. Harte, A. Hastings, P.A. Market, N.D. Martínez, A. Mooers, P. Roopnarine, G. Vermeij, J.W. Williams, R. Gillespie, J. Kitzes, C. Marshall, N. Matzke, D.P. Mindell, E. Revilla and A.B. Smith (2012), 'Approaching a state shift in Earth's biosphere', *Nature*, **486** (7401), 52–8.

Barrett, S. (2011), 'Avoiding disastrous climate change is possible but not inevitable', *Proceedings of the National Academy of Sciences of the United States of America*, **108** (29), 11733–4.

Barrett, S. and A. Dannenberg (2012), 'Climate negotiations under scientific uncertainty', *Proceedings of the National Academy of Sciences of the United States of America*, **109** (43), 17372–6.

Bass, S. (2009), 'Keep off the grass', *Nature Climate Change*, **3** (0910), 113–14.

Bellamy, R., J. Chilvers, N.E. Vaughan and T.M. Lenton (2012), 'A review of climate geoengineering appraisals', *Wiley Interdisciplinary Reviews: Climate Change*, **3** (6), 597–615.

Bellwood, D.R., T.P. Hughes, C. Folke and M. Nyström (2004), 'Confronting the coral reef crisis', *Nature*, **429**, 827–33.

Benford, R.D. and D.A. Snow (2000), 'Framing processes and social movements: An overview and assessment', *Annual Review of Sociology*, **26**, 611–39.

Berkes, F., J. Colding, and C. Folke (eds) (1998), *Navigating Social–Ecological Systems*, Cambridge, UK: Cambridge University Press.

Bernstein, S. and J. Brunnée (2011), 'Options for broader reform of the Institutional Framework for Sustainable Development (IFSD): Structural, legal and financial aspects'. Consultants' Report. Report prepared for the Secretariat of the United Nations Conference on Sustainable Development. Available at http://www.uncsd2012.org/

rio20/index.php?page=view&type=400&nr=211&menu=45 (accessed 18 September 2013).

Bicchetti, D. and N. Maystre (2012), 'The synchronized and long-lasting structural change on commodity markets: Evidence from high frequency data', *Munich Personal RePEc Archive*, MPRA Paper No. 37486. Available at http://mpra.ub.uni-muenchen.de/37486/.

Biello, D. (2012), 'Walking the line: How to identify safe limits for human impacts on the planet', *Scientific American*, 13 June 2012. Available at http://www.sci entificamerican.com/article.cfm?id=do-planetary-boundaries-help-humanity-manage-environmental-impacts (accessed 3 December 2013).

Biermann, F. (2000), 'The case for a world environment organization', *Environment*, **42** (9), 22–31.

Biermann, F. (2006), 'Whose experts? The role of geographic representation in global environmental assessments', in R.B. Mitchell, W.C. Clark, D.W. Cash and N.M. Dickson (eds), *Global Environmental Assessments: Information and Influence*, Cambridge, MA: The MIT Press, pp. 87–112.

Biermann, F. (2007), 'Earth system governance as a crosscutting theme of global change research', *Global Environmental Change*, **17** (3–4), 326–37.

Biermann, F. (2009), 'Environmental policy integration and the architecture of global environmental governance', *International Environmental Agreements*, **9**, 351–69.

Biermann, F. (2012), 'Planetary boundaries and earth system governance: Exploring the links', *Ecological Economics*, **81**, 4–9.

Biermann, F. and S. Bauer (2005), *A World Environment Organization – Solution or Threat for Effective International Environmental Governance?* Aldershot: Ashgate.

Biermann, F. and P. Pattberg (eds) (2012), *Global Environmental Governance Reconsidered*, Cambridge, MA: The MIT Press.

Biermann, F. and B. Siebenhüner (2009), *Managers of Global Change – The Influence of International Environmental Bureaucracies*, Cambridge, MA: The MIT Press.

Biermann, F., K. Abbott, S. Andresen, K. Bäckstrand, S. Bernstein, M.M. Betsill, H. Bulkeley, B. Cashore, J. Clapp, C. Folke, A. Gupta, J. Gupta, P.M. Haas, A. Jordan, N. Kanie, T. Kluvánková-Oravská, L. Lebel, D. Liverman, J. Meadowcroft, R.B. Mitchell, P. Newell, S. Oberthür, L. Olsson, P. Pattberg, R. Sánchez-Rodríguez, H. Schroeder, A. Underdal, S. Camargo Vieira, C. Vogel, O.R. Young, A. Brock and R. Zondervan (2012), 'Navigating the Anthropocene: Improving Earth system governance', *Science*, **335**, 1306–7.

Biermann, F., M.M. Betsill, J. Gupta, N. Kanie, L. Lebel, D. Liverman et al. (2009), 'Earth system governance: People, places and the planet'. Science and Implementation Plan of the Earth System Governance Project. Earth System Governance Report 1, IHDP Report 20 (Bonn, IHDP: The Earth System Governance Project).

Biggs, R., S.R. Carpenter and W.A. Brock (2009), 'Turning back from the brink: Detecting an impending regime shift in time to avert it', *Proceedings of the National Academy of Sciences of the United States of America*, **106** (3), 826–31.

Bipartisan Policy Center (2011), *Geoengineering: A National Strategic Plan for Research on the Potential Effectiveness, Feasibility, and Consequences of Climate Remediation Technologies*, Washington, DC: Bipartisan Policy Center.

Black, R. (2011), 'Climate fix technical test put on hold', *BBC News*. Available at http://www.bbc.co.uk/news/science-environment-15132989 (accessed 14 September 2013).

Blackstock, J. (2012), 'Researchers can't regulate climate engineering alone', *Nature*, **486**, 159.

Blackstock, J., D.S. Battisti, K. Caldeira, D.M. Eardley, J.I. Katz, D.W. Keith, A.A.N. Patrinos, D.P. Schrag, R.H. Socolow and S.E. Koonin (2009), 'Climate engineering responses to climate emergencies', Novim Report, Santa Barbara, California. Archived online at http://arxiv.org/pdf/0907.5140.

Blas, J. (2011), 'High-speed trading blamed for sugar prices'. Available at http://www.ft.com/cms/s/0/05ba0b60-33d8-11e0-b1ed-00144feabdc0.html#axzz1bLsJvJ49 (accessed 10 September 2013).

Blas, J. (2012), 'Commodity market's algorithmic challenge'. Available at http://www.ft.com/intl/cms/s/0/79722992-750f-11e1-90d1-00144feab49a.html#axzz27kV9UZ85 (accessed 10 September 2013).

Blench, M. (2008), 'Global Public Health Intelligence Network', paper presented at the 8th Association for Machine Translation in the Americas (AMTA) Conference, Hawaii, 21–5 October.

Blomqvist, L., T. Nordhaus and M. Shellenberger (2012), *The Planetary Boundaries Hypothesis: A Review of the Evidence*, Oakland, CA: Breakthrough Institute.

Boardman, R. (2010), *Governance of Earth Systems – Science and its Uses*, Hampshire: Palgrave Macmillan.

Bodansky, D. (2011), 'Governing climate engineering: scenarios for analysis'. Discussion Paper 2011-47, Cambridge, MA: Harvard Project on Climate Agreements, September 2011.

Bodin, O. and J. Norberg (2005), 'Information network topologies for enhanced local adaptive management', *Environmental Management*, **35** (2), 175–93.

Bodle, R., G. Homan, S. Schiele and E. Tedsen (2012), *The Regulatory Framework for Climate-Related Geoengineering Relevant to the Convention on Biological Diversity. Part II of: Geoengineering in Relation to the Convention on Biological Diversity: Technical and Regulatory Matters*, Montreal: Secretariat of the Convention on Biological Diversity, Technical Series No. 66, p. 152.

Boettinger, C. and A. Hastings (2012), 'Quantifying limits to detection of early warning for critical transitions', *Journal of the Royal Society Interface*, **9** (75), 2527–39.

Boin, A., P. 't Hart, E. Stern and B. Sundelius (2005), *The Politics of Crisis Management – Public Leadership under Pressure*, Cambridge: Cambridge University Press.

Borin, A. and V. Di Nino (2012), 'The role of financial investments in agricultural commodity derivatives markets', *Banca D'Italia*, Working Paper No. 849.

Brand, S. (2013), 'Opinion: The case for reviving extinct species', *National Geographic News*. Available at http://news.nationalgeographic.com/news/2013/03/130311-deextinction-reviving-extinct-species-opinion-animals-science/ (accessed 24 June 2013).

Brand, R. and J. Fischer (2012), 'Overcoming the technophilia/technophobia split in environmental discourse', *Environmental Politics*, **22** (2), 37–41.

Brashares, J.S., P. Arcese, M.K. Sam, P.B. Coppolillo, A.R.E. Sinclair and A. Balmford (2004), 'Bushmeat hunting, wildlife declines and fish supply in West Africa', *Science*, **306** (5699), 1180–83.

Bravata, D. and K. McDonald (2004), 'Review systematic review: Surveillance systems for early detection of bioterrorism-related diseases', *Annals of Internal Medicine*, **140** (11), 910–24.

Brondizio, E.S., E. Ostrom and O.R. Young (2009), 'Connectivity and the governance of multilevel social–ecological systems: The role of social capital', *Annual Review Environmental Resources*, **34**, 253–78.

Brook, B.W., E.C. Ellis, M.P. Perring, A.W. Mackay and L. Blomqvist (2013), 'Does the terrestrial biosphere have planetary tipping points?' *Trends in Ecology & Evolution*, **2012**, 1–6.

Brownstein, J., C.C. Freifeld and L.C. Madoff (2009), 'Digital disease detection – harnessing the Web for public health surveillance', *New England Journal of Medicine*, **360**, 2153–7.

Bruns, A. (2007), 'Methodologies for mapping the political blogosphere: An exploration using the IssueCrawler research tool', *First Monday*, **12** (5–7).

Burns, W.C.G. and A.L. Strauss (2013), *Climate Change Geoengineering*

  – *Philosophical Perspectives, Legal Issues, and Governance Frameworks*, Cambridge, UK: Cambridge University Press.

Butler, D. (2012), 'Flu surveillance lacking', *Nature*, **483** (7391), 520–22.

Butler, D. and D. Cyranoski (2013), 'Flu papers spark row over credit for data', *Nature*, **497** (7447), 14–15.

Büyükşahin, B., M.S. Haigh and M.A. Robe (2010), 'Commodities and equities: Ever a "Market of One"', *Journal of Alternative Investments*, **12** (3), 76–95.

Cameron, J. and J. Abouchar (1991), 'The precautionary principle: A fundamental principle of law and policy for the protection of the global environment', *Boston College International and Comparative Law Review*, **14** (1), 1–27.

Campbell, John L. (2002), 'Ideas, politics, and public policy', *Annual Review of Sociology*, **28**, 21–38.

Cárdenas, J.C. and E. Ostrom (2004), 'What do people bring into the game? Experiments in the field about cooperation in the commons', *Agricultural Systems*, **82** (3), 307–26.

Carpenter, S.R. and E. Bennett (2011), 'Reconsideration of the planetary boundary for phosphorus', *Environmental Research Letters*, **6** (1), 1–12.

Carpenter, S.R. and W.A. Brock (2004), 'Spatial complexity, resilience and policy diversity: Fishing on lake-rich landscapes', *Ecology and Society*, **9** (1), 8.

Carpenter, S., D. Ludwig and W. Brock (1999), 'Management of eutrophication for lakes subject to potentially irreversible change', *Applied Ecology*, **9**, 751–71.

Cash, D.W., W.C. Clark, F. Alcock, N.M. Dickson, N. Eckley, D.H. Guston, J. Jäger and R.B. Mitchell (2003), 'Knowledge systems for sustainable development', *Proceedings of the National Academy of Sciences*, **100** (14), 8086–91.

Castells, M. (2009), *Communication Power*, Oxford: Oxford University Press.

Chaboud, A., B. Chiquoine, E. Hjalmarsson and C. Vega (2009), 'Rise of the machines: Algorithmic trading in the foreign exchange market', Board of Governors of the Federal Reserve System International Finance, Discussion Papers Number 980 (October 2009).

Chambers, W.B. (2008), *Interlinkages and the Effectiveness of Multilateral Environmental Agreements*, Tokyo: United Nations University Press.

Chan, E.H., T.F. Brewer, L.C. Madoff, M.P. Pollack, A.L. Sonricker, M. Keller, C.C. Freifeld, M. Blench, A. Mawudeku and J.S. Brownstein (2010), 'Global capacity for emerging infectious disease detection', *Proceedings of the National Academy of Sciences*, **107** (50), 21701–6.

Chen, Y., C. Jayaprakash and E. Irwin (2012), 'Threshold management in a coupled economic–ecological system', *Journal of Environmental Economics and Management*, **64** (3), 442–55.

Clapp, J. (2009), 'Food price volatility and vulnerability in the global South: Considering the global economic context', *Third World Quarterly*, **30** (6), 1183–96.

Clapp, J. and M.J. Cohen (eds) (2009), *The Global Food Crisis: Governance Challenges and Opportunities*, Waterloo, CA: Wilfrid Laurier University Press.

Conley, D.J. (2012). 'Ecology: Save the Baltic Sea', *Nature*, **486**, 463–4.

Crépin, A.S. (2007), 'Using fast and slow processes to manage resources with thresholds', *Environmental and Resource Economics*, **36** (2), 191–213.

Cressey, D. (2012), 'Geoengineering experiment cancelled amid patent row', *Nature*, **485**, 429.

Cressey (n.d.), 'Cancelled project spurs debate over geoengineering patents', *Nature* available at http://www.nature.com/news/cancelled-project-spurs-debate-over-geoengineering-patents-1.10690 (accessed 14 September 2013).

Crutzen, P.J. (2002), 'Geology of mankind', *Nature*, **415**, 23.

Dakos, V., S.R. Carpenter, W. Brock, A.M. Ellison, V. Guttal, A.R. Ives, S. Kéfi, V. Levina, D.A. Seekell, E.H. van Nes and M. Scheffer (2012), 'Methods for detecting early warnings of critical transitions in time series illustrated using simulated ecological data', *PloS One*, **7** (7), e41010.

Dalton, R. (2005), 'Conservation policy: Fishy futures', *Nature*, **437**, 473–4.

Daly, H.E. (1990), 'Towards some operational principles of sustainable development', *Ecological Economics*, **2**, 1–6.

De Schutter, O. (2010), 'Food commodities speculation and food price crises: Regulation to reduce the risks of price volatility', Nations Special Rapporteur on the Right to Food, Briefing Note #2.

Dean Moore, K. (2013), 'Anthropocene is the wrong word', *Earth Island Journal*, Spring 2013. Available at http://www.earthisland.org/journal/index.php/eij/article/anthropocene_is_the_wrong_word (accessed 4 September 2013).

DeFries, R.S., F.S. Chapin and J. Syvitski (2012), 'Planetary opportunities: A social contract for global change science to contribute to a sustainable future', *BioScience*, **62** (6), 603–6.

Derocher, A.E., J. Aars, S.C. Amstrup, A. Cutting, N.J. Lunn, P.K. Molnár, M.E. Obbard, E.I. Stirling, G.W. Thiemann, D. Vongraven, O. Wiig and G. York (2013), 'Rapid ecosystem change and polar bear conservation', *Conservation Letters* (in print).

Dessler, A.E., Z. Zhang, and P. Yang (2008), 'Water-vapor climate feedback inferred from climate fluctuations, 2003–2008', *Geophysical Research Letters*, **35**, L20704.

Deutsch, L., S. Gräslund, C. Folke, M. Troell, M. Huitric, N. Kautsky and L. Lebel (2007), 'Feeding aquaculture growth through globalization: Exploitation of marine ecosystems for fishmeal', *Global Environmental Change*, **17** (2), 238–49.

Diamandis, P.H. and S. Kotler (2012), *Abundance – The Future is Better Than You Think*, New York: Free Press.

Dietz, T., E. Ostrom and P.C. Stern (2003), 'The struggle to govern the commons', *Science*, **302** (5652), 1907–12.

Dimitrov, R.S., D.F. Sprinz, G.M. Digiusto and A. Kelle (2007), 'International nonregimes: A research agenda', *International Studies Review*, **9**, 230–58.

Dingwerth, K. and P. Pattberg (2006), 'Global governance as a perspective on world politics', *Global Governance: A review of multilateralism and international organizations*, **12** (2), 185–203.

Dodds, F., K. Schneeberger and F. Ullah (2012), *Review of Implementation of Agenda 21 and the Rio Principles – Synthesis*, New York: United Nations Department of Economic and Social Affairs Division for Sustainable Development.

Doshi, P. (2011), 'The elusive definition of pandemic influenza', *Bulletin of the World Health Organization*, **89** (7), 532–8.

Dry, S. and M. Leach (eds) (2010), *Epidemics: Science, Governance and Social Justice*, London: Earthscan.

Duit, A. and V. Galaz (2008), 'Governing complexity – insights and emerging challenges', *Governance*, **21** (3), 311–35.

Dutt, A.K., R. Akhtar and M. Mcveigh (2006), 'Review: Surat plague of 1994 re-examined', *Southeast Asian Journal of Tropical Medicine and Public Health*, **37** (4), 755–60.

Eakin, H.C. and M.B. Wehbe (2008), 'Linking local vulnerability to system sustainability in a resilience framework: Two cases from Latin America', *Climatic Change*, **93** (3–4), 355–77.

Eakin, H., A. Winkels and J. Sendzimir (2009), 'Nested vulnerability: Exploring cross-scale linkages and vulnerability teleconnections in Mexican and Vietnamese coffee systems', *Environmental Science & Policy*, **12** (4), 398–412.

Ebbesson, J. (2010), 'The rule of law in governance of complex socio-ecological change', *Global Environmental Change*, **20** (3), 414–22.

Economist, The (2012), 'The global environment: Boundary conditions', *The Economist*, 16 June 2012. Available at http://www.economist.com/node/21556897 (accessed 2 December 2013).

Ehrenfeld, D. (2013), 'Resurrected mammoths and dodos? Don't count on it', *Guardian*, 23 March 2013. Available at http://www.theguardian. com/commentisfree/2013/mar/23/de-extinction-efforts-are-waste-of-tim e-money (accessed 18 September 2013).

Elbe, S. (2010), 'Haggling over viruses: the downside risks of securitizing infectious disease', *Health policy and planning*, **25** (6), 476–85.

El-Dukheri, I., N. Elamin and M. Kherallah (2011), 'Farmers' response to soaring food prices in the Arab region', *Food Security*, **3** (S1), 149–62.

Ellis EC. (2011), 'Anthropogenic transformations of the terrestrial biosphere', *Philosophical Transactions of the Royal Society of London*, Series A, **369**, 1010–35.

Ellis, E. (2012), 'The planet of no return – human resilience on an artificial Earth', *Breakthrough Journal*, Winter 2012. Available at http:// thebreakthrough.org/index.php/journal/past-issues/issue-2/the-planet-of-no-return/ (accessed 5 September 2013).

Ellis, E.C. and N. Ramankutty (2008), 'Putting people in the map: anthropogenic biomes of the world', *Frontiers in Ecology and the Environment*, **6**, 439–47.

Elmgren, R. (2012), 'Eutrophication: Political backing to save the Baltic Sea', *Nature*, **487**, 432.

ETC Group (2009), 'ETC Group Submission to Royal Society Working Group on Geo-Engineering' available at http://www.etcgroup.org/fr/ content/etc-group-submission-royal-society-working-group-geo-engine ering-2009 (accessed 19 June 2013).

EU Climate Change Expert Group (2008), *The 2°C target – Background on Impacts, Emission Pathways, Mitigation Options and Costs*, EU Climate Change Expert Group (EG Science).

EurActiv (2011), 'EU wants market-based answer to commodities surge' available at http://www.euractiv.com/sustainability/eu-wants-market-based-answer-com-news-502927 (accessed 10 September 2013).

EurActiv (2011b), 'EU wants to tame "financialisation" of commodity markets'. Available at http://www.euractiv.com/specialreport-raw materials/eu-wants-tame-financialisation-c-news-502734 (accessed 10 September 2013).

European Commission (2011), 'Tackling the challenges in commodity markets and on raw materials', European Commission, Brussels. COM(2011) 25 final. Available at http://eur-lex.europa.eu/LexUriServ/ LexUriServ.do?uri=COM:2011:0025:FIN:en:PDF (accessed 9 October 2013).

European Environment Agency (2013), 'Late lessons from early warnings: Science, precaution, innovation', European Environment Agency, Copenhagen. Report No 1/2013.

Eurosurveillance (2002), 'Outbreak of influenza, Madagascar, July–August 2002', **7** (12).

Fairhead, J., M. Leach and I. Scoones (2012), 'Green grabbing: A new appropriation of nature', *Journal of Peasant Studies* (July), 37–41.

Faunce, T.A. (2012), *Nanotechnology for a Sustainable World: Global artificial photosynthesis as the moral culmination of nanotechnology*, Cheltenham, UK and Northampton, MA, USA: Edward Elgar Publishing.

Fidelman, P., L. Evans, M. Fabinyi, S. Foale, J. Cinner and F. Rosen (2012), 'Governing large-scale marine commons: Contextual challenges in the coral triangle', *Marine Policy*, **36** (1), 42–53.

Fidler, D.P. (2004), SARS, *Governance and the Globalization of Disease*, Houndmills: Palgrave Macmillan.

Fidler D.P. (2008), 'Influenza virus samples, international law, and global health diplomacy', *Emerging Infectious Diseases*, **14** (1), 88–94.

Fischer, J., D.B. Lindenmayer and A.D. Manning (2006), 'Biodiversity, ecosystem function, and resilience: Ten guiding principles for commodity production landscapes', *Frontiers in Ecology and the Environment*, **4** (2), 80–86.

Fleming, J.R. (2010), *Fixing the Sky: The checkered history of weather and climate control*, New York, USA: Columbia University Press.

Folke, C., T. Hahn, P. Olsson and J. Norberg (2005), 'Adaptive governance of social–ecological systems', *Annual Review of Environmental Resources*, **30**, 441–73.

Folke, C., S.R. Carpenter, B. Walker, M. Scheffer, T. Chapin and J. Rockström (2010), 'Resilience thinking: Integrating resilience, adaptability, and transformability', *Ecology and Society*, **15** (4), 20.

Folke, C., S. Carpenter, B. Walker, M. Scheffer, T. Elmqvist, L. Gunderson and C.S. Holling (2004), 'Regime shifts, resilience and biodiversity in ecosystem management', *Annual Review of Ecology Evolution and Systematics*, **35**, 557–81.

Folke, C., Å. Jansson, J. Rockström, P. Olsson, S.R. Carpenter, F.S. Chapin III, A-S. Crépin, G. Daily, K. Danell, J. Ebbesson, T. Elmqvist, V. Galaz, F. Moberg, M. Nilsson, H. Österblom, E. Ostrom, Å. Persson, G. Peterson, S. Polasky, W. Steffen, B. Walker and F. Westley (2011), 'Reconnecting to the biosphere', *Ambio*, **40**, 719–38.

Food and Agricultural Organization (FAO) (2009), 'Fisheries and aquaculture Technical Paper', Fisheries and Aquaculture Dept., no. 530, Rome: FAO, p. 217.

Food and Agricultural Organization (FAO) (2010), 'Principles for responsible agricultural investment that respects rights, livelihoods and resources' (Extended Version), Rome: FAO.

Food and Agricultural Organization (FAO) (2011a), 'The state of food insecurity in the world 2011 – how does international price volatility affect domestic economies and food security?', Rome: FAO/IFAD/ WFP.

Food and Agricultural Organization (FAO) (2011b), 'Price volatility in food and agricultural markets: policy responses'. Policy Report by FAO, IFAD, IMF, OECD, UNCTAD, WFP, the World Bank, the WTO, IFPRI and the UN HLTF (2 June 2011).

Forster, J., A.G. Hirst and D. Atkinson (2012), 'Warming-induced reductions in body size are greater in aquatic than terrestrial species', *PNAS*, **109** (47), 19310–14.

Forster, P. (2012), *To Pandemic or Not? Reconfiguring Global Responses to Influenza*, STEPS Working Paper 51, Brighton: STEPS Centre.

Foster, K.R., P. Vecchia and M.H. Repacholi (2000), 'Science and the precautionary principle', *Science*, **288** (5468), 979–81.

Friedrich, J. (2013), *Modeling for Planetary Boundaries – A network analysis of the representations of complex human–environmental interactions in integrated global models*. Thesis work, Stockholm Resilience Centre, Stockholm University.

Galaz, V. (2009), 'Pandemic 2.0: Can information technology help save the planet?', *Environment*, **51** (6), 20–28.

Galaz, V. (2011), 'Double complexity – information technology and reconfigurations in adaptive governance', in E. Boyd and C. Folke (eds), *Adapting Institutions – Governance, Complexity and Social–Ecological Resilience*, Cambridge, UK: Cambridge University Press, pp. 193–215.

Galaz, V. (2012). 'Geo-engineering, governance and social–ecological systems', *Ecology and Society*, **17** (1), 24.

Galaz, V., S. Cornell and J. Rockström (2012), 'Planetary boundaries concept is valuable', *Nature*, **486**, 191 (14 June 2012).

Galaz, V., F. Biermann, C. Folke, M. Nilsson and P. Olsson (2012), 'Global environmental governance and planetary boundaries: An introduction', *Ecological Economics*, **18**, 9–11.

Galaz, V., B. Crona, H. Österblom, P. Olsson and C. Folke (2011), 'Polycentric systems and interacting planetary boundaries: Emerging governance of climate change – ocean acidification – marine biodiversity', *Ecological Economics*, **81**, 21–32.

Galaz, V., J. Gars, F. Moberg, B. Nyqvist and J. Lundberg (2013), 'The ultimate disconnect from the biosphere? Why ecologists should care about financial markets and algorithmic trade', Working Paper, Stockholm Resilience Centre, Stockholm University. Available at https:// docs.google.com/file/d/0B0H9KDhnWEq-cHRnNmZKZWc2cnM/edit ?usp=sharing (accessed 19 September 2013).

Galaz, V., F. Moberg, E-K. Olsson, E. Paglia and C. Parker (2010), 'Institutional and political leadership dimensions of cascading ecological crises', *Public Administration*, **89** (2), 361–80.

Galaz, V., P. Olsson, T. Hahn, C. Folke and U. Svedin (2008), 'The problem of fit among biophysical systems, environmental and resource regimes, and broader governance systems: Insights and emerging challenges', in O.R. Young, L.A. King and H. Schroeder (eds), *Institutions and Environmental Change – Principal Findings, Applications, and Research Frontiers*, Cambridge, MA: The MIT Press, pp. 147–82.

Galaz, V., B. Crona, T. Daw, Ö. Bodin, M. Nyström and P. Olsson (2010), 'Can web crawlers revolutionize ecological monitoring?', *Frontiers in Ecology and the Environment*, **8** (2), 99–104.

Galaz, V., F. Biermann, B. Crona, D. Loorbach, C. Folke, P. Olsson et al. (2012), 'Planetary boundaries – exploring the challenges for global environmental governance', *Current Opinion in Environmental Sustainability*, **4** (1), 80–87.

Gardiner, S. (2006), 'A core precautionary principle', *Journal of Political Philosophy*, **14** (1), 33–60.

Garrett, L. (1996), 'The return of infectious disease', *Foreign Affairs*, **75** (1), 66–79.

Garten, R.J., C.T. Davis, C. Russell, B. Shu, S. Lindstrom, A. Balish, W.M. Sessions, X. Xu, E. Skepner, V. Deyde, M. Okomo-Adhiambo, L. Gubareva, J. Barnes, C.B. Smith, S.L. Emery, M.J. Hillman, P. Rivailler, J. Smagala, M. de Graaf, D.F. Burke, R.A.M. Foucheier, C. Pappas, C.M. Alpuche-Aranda, H. López-Gatell, H. Olivera, I. López, C.A. Myers, D. Faix, P.J. Blair, C. Yu, K.M. Keene, P.D. Dotson Jr., D. Boxrud, A.R. Sambol, S.H. Abid, K. St. George, T. Bannerman, A.L. Moore, D.J. Stringer, P. Blevins, G.J. Demmler-Harrison, M. Ginsberg, P. Kriner, S. Waterman, S. Smole, H.F. Guevara, E.A. Belongia, P.A. Clark, S.T. Beatrice, R. Donis, J. Katz, L. Finelli, C.B. Bridges, M. Shaw, D.B. Jernigan, T.M. Uyeki, S.J. Smith, A.I. Klimov and N.J. Cox (2009), 'Antigenic and genetic characteristics of swine-origin 2009 A (H1N1) influenza viruses circulating in humans', *Science*, **325** (5937), 197–201.

Gehring, T. and S. Oberthür (2008), 'Interplay: Exploring institutional interaction', In O.R. Young, L.A. King and H. Schroeder (eds), *Institutions and Environmental Change – Principles Findings, Applications, and Research Frontiers*, Cambridge, MA: The MIT Press, pp. 187–223.

Gehring, T. and S. Oberthür (2009), 'The causal mechanism of interaction between international institutions', *European Journal of International Relations*, **15** (1), 125–56.

Gelcich, S., T.P. Hughes, P. Olsson, C. Folke, O. Defeo, M. Fernández, S. Foale, L.H. Gunderson, C. Rodríguez-Sickert, M. Scheffer, R.S. Steneck and J.C. Castilla (2010), 'Navigating transformations in governance of Chilean marine coastal resources', *Proceedings of the National Academy of Sciences*, **107** (39), 16794–9.

Geman, H. (2005), *Commodities and Commodity Derivatives – Modeling and Pricing for Agriculturals, Metals and Energy*, Chichester, West Sussex: Wiley & Sons.

Gifford, R. (2011), 'The dragons of inaction: Psychological barriers that limit climate change mitigation and adaptation', *The American Psychologist*, **66** (4), 290–302.

Gilbert, C.L. and C.W. Morgan (2010), 'Food price volatility', *Philosophical Transactions of the Royal Society of London*, Series B, Biological sciences, **365** (1554), 3023–34.

Girardi, D. (2012), 'Do financial investors affect the price of wheat?', *PSL Quarterly Review*, **65**, 79–109.

Goldenfeld, N. and L.P. Kadanoff (1999), 'Simple lessons from complexity', *Science*, **284** (5411), 87–9.

Gomber, P., B. Arndt, M. Lutat and T. Uhle (2011), 'High-frequency trading', Goethe Universität Frankfurt am Main, *SSRN Electronic Journal*. Available at http://ssrn.com/abstract=1858626.

Goodell, J. (2010), *How to Cool the Planet: Geoengineering and the Audacious Quest to Fix Earth's Climate*, Boston, New York: Houghton Mifflin Harcourt.

Gorton, G. and K.G. Rouwenhorst (2006), 'Facts and fantasies about commodity futures', *The National Bureau of Economic Research (NBER)*, Working Paper No. 10595 (March).

Government Accountability Office (GAO) (2010), 'Climate change: Preliminary observations on geoengineering science, federal efforts, and governance issues', Report GAO-10-546T, Washington, DC: GAO.

Government Accountability Office (GAO) (2011), 'Climate engineering: technical status, future directions, and potential responses', Report GAO-11-71, Washington, DC: GAO.

Government Office for Science (GOS) (2011), 'The future of computer trading in financial markets', Working Paper by UK Government Office for Science, Foresight.

Gowers, T. and M. Nielsen (2009), 'Massively collaborative mathematics', *Nature*, **461**, 879–81.

Grace, D., F. Mutua, P. Ochungo, R. Kruska, K. Jones, L. Brierley, L. Lapar, M. Said, M. Herrero, M.P. Phuc, N.B. Thao, I. Akuku and F. Ogutu (2012), *Mapping of Poverty and Likely Zoonoses Hotspots*, Nairobi: International Livestock Research Institute (ILRI).

Greenfeld, K.T. (2006), *China Syndrome: The True Story of the 21st Century's First Great Epidemic*, New York: Harper Collins.

Griggs, D., M. Stafford-Smith, O. Gaffney, J. Rockström, M.C. Öhman, P. Shyamsundar, W. Steffen, G. Glaser, N. Kanie and I. Noble (2013), 'Policy: Sustainable development goals for people and planet', *Nature*, **495** (7441), 305–7.

Groffman, P.M., J.S. Baron, T. Blett, A.J. Gold, I. Goodman, L.H. Gunderson, B.M. Levinson, M.A. Palmer, H.W. Paerl, G.D. Peterson, N. LeRoy Poff, D.W. Rejeski, J.F. Reynolds, M.G. Turner, K.C. Weathers and J. Wiens (2006), 'Ecological thresholds: The key to successful environmental management or an important concept with no practical application?', *Ecosystems*, **9** (1), 1–13.

Grossman, S.J. (1977), 'The existence of futures markets, noisy rational expectations and informational externalities', *Review of Economic Studies*, **44** (3), 431–49.

Gubler, D.J. (1998), 'Resurgent vector-borne diseases as a global health problem', *Emerging Infectious Diseases*, **4** (3), 442–50.

Guimerà, R., B. Uzzi, J. Spiro and L.A. Nunes Amaral (2005), 'Team assembly mechanisms determine collaboration network structure and team performance', *Science*, **308**, 697–702.

Gunderson, L. (1999), 'Resilience, flexibility and adaptive management – antidotes for spurious certitude?', *Conservation Ecology*, **3** (1), 7.

Gunderson, L.H. and C.S. Holling (2002), *Panarchy – Understanding Transformations in Systems of Humans and Nature*, Washington, DC: Island Press.

Gurevich, A. and N. Borisov (1995), 'Artificial ozone layer', *Physics Letters A*, **9601** (November), 281–8. Available at http://www.science direct.com/science/article/pii/0375960195006905.

Haas, P.M. (1992), 'Introduction: Epistemic communities and international policy coordination', *International Organization*, **46** (1), 1–35.

Hamilton, C. (2012), 'The philosophy of geoengineering', Working Paper to the IMPLICC symposium 'The atmospheric science and economics of climate engineering via aerosol injection' held at the Max Planck Institute for Chemistry, Mainz, Germany, 14–16 May, 2012. Available at http://www.clivehamilton.com/papers/philosophy-of-geo engineering/ (accessed 3 September 2013).

Hamilton, C. (2012), *Earth Masters: The Dawn of the Age of Climate Engineering*. New Haven and London: Yale University Press.

Harmon, D., B. Stacey and Y. Bar-Yam (2010), 'Networks of economic market interdependence and systemic risk', arXiv preprint, NECSI Report 2009-03-01, arXiv:1011.3707.

Harremoës P, D. Gee, M. Macgarvin et al. (2001), *Late Lessons from*

Early Warnings: The Precautionary Principle 1896–2000. Copenhagen: European Environment Agency.

Heckbert, S., T. Baynes and A. Reeson (2010), 'Agent-based modeling in ecological economics', *Annals of the New York Academy of Sciences*, **1185**, 39–53.

Helbing, D. (2013), 'Globally networked risks and how to respond', *Nature*, **497** (7447), 51–9.

Heller, N.E. and E.S. Zavaleta (2009), 'Biodiversity management in the face of climate change: A review of 22 years of recommendations', *Biological Conservation*, **142** (1), 14–32.

Hendershott, T., C.M. Jones and A.J. Menkveld (2011), 'Does algorithmic trading improve liquidity?', *The Journal of Finance*, **LXVI** (1), 1–33.

Henriques, D.B. (2008), 'Price volatility adds to worry on US farms'. Available at http://www.nytimes.com/2008/04/22/business/22commodity.html?pagewanted=all&_r=0 (accessed 10 September 2013).

Hewitt, N., N. Klenk, L. Smith, D.R. Bazely, N. Yan, S. Wood, J.I MacLellan, C. Lipsig-Mumme and I. Henriques (2011), 'Taking stock of the assisted migration debate', *Biological Conservation*, **144** (11), 2560–72.

Heylighen, F. (1999), 'Collective intelligence and its implementation on the Web: Algorithms to develop a collective mental map', *Computational and Mathematical Organization Theory*, **5** (3), 253–80.

Heymann, D.L. (2003), 'The evolving infectious disease threat: Implications for national and global security', *Journal of Human Development*, **4** (2), 191–206.

Heymann, D. (2006), 'SARS and emerging infectious diseases: A challenge to place global solidarity above national sovereignty', *Annals Academy of Medicine Singapore*, **35** (5), 350–53.

Heyward, C. (2013), 'Situating and abandoning geoengineering: A typology of five responses to dangerous climate change', *Political Science and Politics*, **46** (01), 23–7.

High Frequency Traders (2012), 'SEC votes in favour of high frequency trading audits' available at http://highfrequencytraders.com/news/2379/sec-votes-favour-high-frequency-trading-audits (accessed 10 September 2013).

Hobbs, R.J. et al. (2009), 'Novel ecosystems: Implications for conservation and restoration', *Trends in Ecology and Evolution*, **24**, 599–605.

Hoegh-Guldberg, O., L. Hughes, S. MacIntyre, D.B. Lindenmayer, C. Parmesan, H.P. Possingham and C.D. Thomas (2008), 'Assisted colonization and rapid climate change', *Science*, **321**, 345–6.

Hoegh-Guldberg, O., P.J. Mumbym, A.J. Hooten, R.S. Steneck et al.

(2007), 'Coral reefs under rapid climate change and ocean acidification', *Science*, **318** (5857), 1737–42.

Holling, C.S. (1973), 'Resilience and stability of ecological systems', *Annual Review of Ecology and Systematics*, **4**, 1–23.

Holling, C.S. (2004), 'From complex regions to complex worlds', *Ecology and Society*, **9** (1), 11.

Holling, C.S. and G.K. Meffe (1996), 'Command and control and the pathology of natural resource management', *Conservation Biology*, **10** (2), 328–37.

Homer-Dixon, T. (2002), *The Ingenuity Gap – Facing the Economic, Environmental and other Challenges of an Increasingly Complex and Unpredictable World*, New York: Vintage Books.

Hooghe, L. and G. Marks (2003), 'Unravelling the central state, but how? Types of multi-level governance', *American Political Science Review*, **97**, 233–43.

Horan, R.D., E.P. Fenichel, K.L.S. Drury and D.M. Lodge (2011), 'Managing ecological thresholds in coupled environmental–human systems', *Proceedings of the National Academy of Sciences of the United States of America*, **108** (18), 7333–8.

House of Commons (2010), 'The regulation of geoengineering – fifth report of session 2009–10', Report, together with formal minutes, oral and written evidence, London, UK: House of Commons Science and Technology Committee.

Hsu V.P., M.J. Hossain, U.D. Parashar, M.M. Ali, T.G. Ksiazek, I. Kuzmin, M. Niezgoda, C. Rupprecht, J. Bresee and R.F. Breiman (2004), 'Nipah virus encephalitis reemergence, Bangladesh', *Emerging Infectious Disease*, **10** (12), 2082–7.

Hughes, T.P., C. Linares, V. Dakos, I. van de Leemput and E.H. van Nes (2013), 'Living dangerously on borrowed time during slow, unrecognized regime shifts', *Trends in Ecology and Evolution*, **28** (3), 149–55.

Hulme, M. (2012), 'On the "two degrees" climate policy target', in O. Edenhofer, J. Wallacher, H. Lotze-Campen, M. Reder, B. Knopf and J. Müller (eds), *Climate Change, Justice and Sustainability: Linking Climate and Development Policy*, Dordrecht, Netherlands: Springer, pp. 122–5.

Institute of Medicine (IOM) (2008), *Review of the DOD-GEIS Influenza programs: Strengthening global surveillance and response*, Washington, DC: The National Academies Press.

Institute of Medicine and National Research Council (2008), 'Achieving sustainable global capacity for surveillance and response to emerging diseases of zoonotic origin', Workshop Report, Washington, DC: The National Academies Press.

Irwin, S.H. and D.R. Sanders (2011), 'Index funds, financialization, and commodity futures markets', *Applied Economic Perspectives and Policy*, **33** (1), 1–31.

Jager, W., M.A. Janssen and C.A.J. Vlek (2002), 'How uncertainty stimulates over-harvesting in a resource dilemma: Three process explanations', *Journal of Environmental Psychology*, **22**, 247–63.

Janssen, M.A., J.M. Anderies and B.H. Walker (2004), 'Robust strategies for managing rangelands with multiple stable attractors', *Journal of Environmental Economics and Management*, **47** (1), 140–62.

Janssen, M.A., R. Holahan, A. Lee and E. Ostrom (2010), 'Lab experiments for the study of social–ecological systems', *Science*, **328** (5978), 613–17.

Jenkins, H. (2006), *Convergence Culture – Where Old and New Media Collide*, New York: New York University Press.

Jensen, K. (2002), 'The moral foundation of the precautionary principle', *Journal of Agricultural and Environmental Ethics*, **15** (1), 39–55.

Jervis. R. (1997), *System Effects: Complexity in Political and Social Life*, Princeton: Princeton University Press.

Jewert, J. (2013), 'Utopiska krav för Östersjön', *Dagens Nyheter*, 2013-06-11. Available at http://www.dn.se/ledare/kolumner/utopiska-krav-for-ostersjon/ (accessed 5 September 2013).

Jewert, J. (2013), 'Politiken har gått vilse', *Forskning och Framsteg*, **7**. Available at http://fof.se/tidning/2013/7/artikel/kronika-politiken-har-gatt-vilse (accessed 5 September 2013).

Johnson, N., G. Zhao, E. Hunsader, H. Qi, N. Johnson, J. Meng and B. Tivnan (2013), 'Abrupt rise of new machine ecology beyond human response time', *Scientific Reports*, **3**, 2627. doi:10.1038/srep02 627.

Jones, B.A, D. Grace, R. Kock, S. Alonso, J. Rushton, M.Y. Said, D. McKeever, F. Mutua, J. Young, J. McDermott and D. Udo Pfeiffer (2013), 'Zoonosis emergence linked to agricultural intensification and environmental change', *Proceedings of the National Academy of Sciences of the United States of America*, **110** (21), 8399–404.

Jones, K.E., N. Patel, M.A. Levy, A. Storeygard, D. Balk, J.L. Gittleman and P. Daszak (2008), 'Global trends in human emerging infectious diseases', *Nature*, **451**, 990–93.

Kanie, N., M.M. Betsill, R. Zondervan et al. (2012), 'A charter moment: Restructuring governance for sustainability', *Public Administration and Development*, **32**, 292–304.

Kaufmann, D. (2003), *Rethinking Governance: Empirical lessons challenge orthodoxy*, Washington, DC: World Bank.

Keith, D.W. (2010), 'Photophoretic levitation of engineered aerosols for geoengineering', *Proceedings of the National Academy of Sciences*, **107** (38), 16428–31.

Kelly, K. (2010), *What Technology Wants*, New York: Penguin.

Kim, R.E. (2013), 'The emergent network structure of the multilateral environmental agreement system', *Global Environmental Change*, preprint e-view http://dx.doi.org/10.1016/j.gloenvcha.2013.07.006.

Kim, R.E. and K. Bosselmann (2013), 'International environmental law in the Anthropocene: Towards a purposive system of multilateral environmental agreements', *Transnational Environmental Law*, **2** (02), 285–309.

Kintisch, E. (2010), 'Bill Gates funding geoengineering research', *ScienceInsider*. Available at http://news.sciencemag.org/sciencein sider/2010/01/bill-gates-fund.html (accessed 17 September 2013).

Kinzig, A.P., P. Ryan, M. Etienne, H. Allison, T. Elmqvist and B.H. Walker (2006), 'Resilience and regime shifts: Assessing cascading effects', *Ecology and Society*, **11** (1), 20.

Knopf, B., C. Flachsland and O. Edenhofer (2012), 'The 2C target reconsidered', in O. Edenhofer, J. Wallacher, H. Lotze-Campen, M. Reder, B. Knopf and J. Müller (eds), *Climate Change, Justice and Sustainability: Linking Climate and Development Policy*, Dordrecht, Netherlands: Springer, pp. 121–37.

Koh, L.P. and D.S. Wilcove (2007), 'Cashing in palm oil for conservation', *Nature*, **448** (7157), 993–4.

Kolbert, E. (2010), 'The Anthropocene debate: Marking humanity's impact', in *Yale Environment 360*. Available at http://e360.yale.edu/content/feature.msp?id=2274 (accessed 8 November 2012).

Kooiman, J. (2003), *Governing as Governance*, London: Sage Publications.

Koppenjan, J. and E-H. Klijn (2004), *Managing Uncertainties in Networks: A Network Approach to Problem Solving and Decision Making*, New York: Routledge.

Kosoy, N. and E. Corbera (2010), 'Payments for ecosystem services as commodity fetishism', *Ecological Economics*, **69**, 1228–36.

Krasner, S.D. (1984), 'Approaches to the State. Alternative conceptions and historical dynamics', *Comparative Politics*, **16** (2), 223–46.

Krasner, S.D. (1991), 'Global communications and national power: Life on the pareto frontier', *World Politics*, **43** (3), 336–66.

Kriebel, D., J. Tickner and P. Epstein (2001), 'The precautionary principle in environmental science', *Environmental Health Perspectives*, **109** (9), 871–6.

Kroszner, R. (2000), 'The economics and politics of financial modernization', *Economic Policy Review*, **6** (4), 25–37.

Kurland, N.B. and T.D. Egan (1999), 'Telecommuting: Justice and control in the virtual organization', *Organization Science*, **10** (4), 500–13.

Lacey, M. (2008), 'Across globe, hunger brings rising anger' 18 April. Available at http://nyti.ms/11LN7fQ (accessed 11 September 2013).

Lambert, E. (2011), *The Futures – The Rise of the Speculator and the Origins of the World's Biggest Markets*, New York: Basic Books.

Lambin, E.F. (2005), 'Conditions for sustainability of human–environment systems: information, motivation and capacity', *Global Environmental Change*, **15** (3), 177–80.

Lambin, E.F. and P. Meyfroidt (2011), 'Global land use change, economic globalization, and the looming land scarcity', *Proceedings of the National Academy of Sciences of the United States of America*, **108** (9), 3465–72.

Lambin, E.F., B.L. Turner, H.J. Geist, S.B. Agbola, A. Angelsen, C. Folke, J. Bruce, O.T. Coomes, R. Dirzog, G. Fischer, P.S. George, K. Homewood, J. Imbernon, R. Leemans, X. Li, E.F. Moran, M. Mortimore, P.S. Ramakrishnan, J.F. Richards, H. Skånes, W. Steffen, G.D. Stone, U. Svedin, T.A. Veldkamp, C. Vogel and J. Xu (2001), 'The causes of land-use and land-cover change: Moving beyond the myths', *Global Environmental Change*, **11** (4), 261–9.

Larigauderie, A. and H.A. Monney (2010), 'The intergovernmental science–policy platform on biodiversity and ecosystem services: Moving a step closer to an IPCC-like mechanism for biodiversity', *Current Opinion in Environmental Sustainability*, **2**, 1–6.

Leach, M. (2013), 'Pathways to sustainability: Building political strategies', in Worldwatch Institute, *State of the World 2013: Is Sustainability Still Possible?*, Washington, DC: Island Press, pp. 234–43.

Leach, M. and I. Scoones (2013), 'The social and political lives of zoonotic disease models: Narratives, science and policy', *Social Science & Medicine*, **88**, 10–17.

Leach, M., I. Scoones and A. Stirling (2010), 'Governing epidemics in an age of complexity: Narratives, politics and pathways to sustainability', *Global Environmental Change*, **20** (3), 369–77.

Leach, M., J. Rockström, P. Raskin, I. Scoones, A.C. Stirling, A. Smith, J. Thompson, E. Millstone, A. Ely, E. Arond, C. Folke and P. Olsson (2012), 'Transforming innovation for sustainability', *Ecology and Society*, **17** (2), 11.

Lenglet, M. (2011), 'Conflicting codes and codings: How algorithmic trading is reshaping financial regulation', *Theory, Culture & Society*, **28** (6), 44–66.

Lenton, Timothy M. (2013), 'Can emergency geoengineering really prevent climate tipping points?', Opinion Article, Geoengineering Our Climate Working Paper and Opinion Article Series. Available at http://wp.me/p2zsRk-7Z (accessed 12 September 2013).

Lenton, T.M. and H.T.P. Williams (2013), 'On the origins of planetary-scale tipping points', *Trends in Ecology and Evolution*, **28** (7), 380–83.

Lenton, T.M., H. Held, E. Kriegler, J.W. Hall, W. Lucht, S. Rahmstorf and H.J. Schellnhuber (2008), 'Tipping elements on the Earth's climate system', *Proceedings of the National Academy of Sciences USA*, **105** (6), 1786–93.

Levy, P. (1999), *Collective Intelligence: Mankind's Emerging World in Cyberspace*, United States: Perseus Books.

Lewis, S. (2012), 'We must set planetary boundaries wisely', *Nature*, **485**, 417.

Lin, A.C. (2009), 'Geoengineering governance', *Issues in Legal Scholarship*, **8** (3), article 2. Available at http://www.bepress.com/ils/vol8/iss3/art2.

Lin, S. (1995), 'Geopolitics communicable diseases: Plague in Surat, 1994', *Economic and Political Weekly*, **30** (46), 2912–14.

Lindahl, T., A-S. Crépin and C. Schill (2012), 'Managing resources with potential regime shifts: Using experiments to explore social–ecological linkages in common resource systems', Beijer Discussion Paper, The Beijer Institute of Ecological Economics, The Royal Swedish Academy of Sciences.

Lindsey, D., P. Cheney and G. Kasper (1990), 'TELCOT: Application of information technology for competitive advantage in the cotton industry', *MIS Quarterly*, **14** (4), 347–57.

Liu, J., T. Dietz, S.R. Carpenter, M. Alberti, C. Folke, E. Moran, A.N. Pell, P. Deadman, T. Kratz, J. Lubchenco, E. Ostrom, Z. Ouyang, W. Provencher, C.L. Redman, S.H. Schneider, and W.W. Taylor (2007), 'Complexity of coupled human and natural systems', *Science*, **317** (5844), 1513–16.

Liu, J., V. Hull, M. Batistella, R. DeFries, T. Dietz, F. Fu, T.W. Hertel, R.C. Izaurralde, E.F. Lambin, S. Li, L.A. Martinelli, W.J. McConnell, E.F. Moran, R. Naylor, Z. Ouyang, K.R. Polenske, A. Reenberg, G. de Miranda Rocha, C.S. Simmons, P.H. Verburg, P.M. Vitousek, F. Zhang, and C. Zhu (2013), 'Framing sustainability in a telecoupled world', *Ecology and Society*, **18** (2), 26.

Loder, A. (2010), 'CFTC to scrutinize algorithmic trading after may market plunge' available at http://www.bloomberg.com/news/2010-10-12/commodity-regulator-to-review-algorithmic-trading-after-may-market-plunge.html (accessed 10 September 2013).

Loorbach, D. (2010), 'Transition management for sustainable development: A prescriptive, complexity-based governance framework', *Governance*, **23** (1), 161–83.

Low, B., E. Ostrom, C. Simon and J. Wilson (2003), 'Redundancy and diversity: Do they influence optimal management?', in F.J. Berkes, J.

Colding and C. Folke (eds), *Navigating Social–Ecological Systems – Building Resilience for Complexity and Change*, pp. 83–113.

Luke, D. and K. Stamatakis (2012), 'Systems science methods in public health: Dynamics, networks, and agents', *Annual Review of Public Health*, **33**, 357–76.

Lynam, T. and K. Brown (2011), 'Mental models in human–environment interactions: theory, policy implications, and methodological explorations', *Ecology and Society*, **17** (3), 24.

Lynas, M. (2011), *The God Species: How the planet can survive the age of humans*, London: Harper Collins.

MacMynowski, D.G., D.W. Keith, K. Caldeira and H.-J. Shin (2011), 'Can we test geoengineering?', *Energy & Environmental Science*, **4** (12), 5044–52.

Madoff, L. and J.P. Woodall (2005), 'The Internet and the global monitoring of emerging diseases: Lessons from the first 10 years of ProMED-mail', *Archives of Medical Research*, **36**, 724–30.

Mäler, K.G., A. Xepapadeas and A. De Zeeuw (2003), 'The economics of shallow lakes', *Environmental and Resource Economics*, **26** (4), 603–24.

Mandel, J.T., C.J. Donlan and J. Armstrong (2010), 'A derivative approach to endangered species conservation', *Frontiers in Ecology and the Environment*, **8** (1), 44–9.

Markus, M.L, B Manville and C.E. Agres (2000), 'What makes a virtual organization work?', *Sloan Management Review*, 13–26.

Marris, E. (2008), 'Moving on assisted migration', *Nature Reports Climate Change*, **2** (0809), 112–13.

Marshall, D. (2002). 'An organization for the world environment: Three models and analysis', *Georgetown International Environmental Law Review*, **15**, 79–103.

May, R.M., S.A. Levin and G. Sugihara (2008), 'Complex systems: Ecology for bankers', *Nature*, **451** (February), 893–5.

McGinnis, M.D. (2000), *Polycentric Games and Institutions*, Ann Arbor, MI: University of Michigan Press.

McGinnis, M.D. (2005), 'Cost and challenges of polycentric governance', Paper for Workshop on Analyzing Problems of Polycentric Governance in the Growing EU, Humboldt University, Berlin, 16–17 June.

McGowan, M. (2010), 'Rise of computerized high frequency trading: Use and controversy', *Duke Law and Technology Review*, **16**, 1–24.

McKelvey, B. (1999), 'Complexity theory in organization science: Seizing the promise or becoming a fad?', *Emergence*, **1** (1), 5–32.

McLachlan, J.S., J.J. Hellmann and M.W. Schwartz (2007), 'A framework for debate of assisted migration in an era of climate change',

Conservation Biology, the Journal of the Society for Conservation Biology, **21** (2), 297–302.

McNally, R. (2005), 'Sociomics! Using the IssueCrawler to map, monitor and engage with the global proteomics research network', *Proteomics*, **5**, 3010–16.

Michelson, E.S. (2005), 'Dodging a bullet: WHO, SARS, and the successful management of infectious disease', *Bulletin of Science, Technology & Society*, **25** (5), 379–86.

Milinski, M., R.D. Sommerfeld, H-H. Krambeck, F.A. Reed and J. Marotzke (2008), 'The collective risk social dilemma and the prevention of simulated dangerous climate change', *Proceedings of the National Academy of Sciences*, **105** (7), 2291–4.

Milkoreit, M. (2013), *Mindmade Politics: The Role of Cognition in Global Climate Change Governance*. Ph.D. Thesis presented to the University of Waterloo in fulfillment of the thesis requirement for the degree of Doctor of Philosophy in Global Governance.

Millennium Ecosystem Assessment (2005), *Ecosystems and Human Well-Being*, Washington, DC: Island Press.

Miller, C. (2006), 'Democratization, international knowledge institutions, and global governance', *Governance*, **20**, 325–57.

Miller, L.M., F. Gans and A. Kleidon (2011), 'Estimating maximum global land surface wind power extractability and associated climatic consequences', *Earth System Dynamics*, **2**, 1–12.

Mirvich, D. (2011), 'The Hathaway effect: How Anne gives Warren Buffett a rise'. Available at http://www.huffingtonpost.com/dan-mirvish/the-hathaway-effect-how-a_b_830041.html (accessed 10 September 2013).

Mitchell, R. (1998), 'Sources of transparency: Information systems in international regimes', *International Studies Quarterly*, **42** (1), 109–30.

Mitchell, R.B., W.C. Clark, D.W. Cash and N.M. Dickson (2006), *Global Environmental Assessments: Information and Influence*, Cambridge, MA: The MIT Press.

Mitra, G., D. DiBartolomeo, A. Banerjee and X. Yu (2011), *Automated Analysis of News to Compute Market Sentiment: Its Impact on Liquidity and Trading*, London: Government Office for Science, Foresight.

Molden, D. (2009), 'Planetary boundaries: The devil is in the detail', *Nature Reports Climate Change*, **3** (0910), 116–17.

Molina, M.J. (2009), 'Planetary boundaries: Identifying abrupt change', *Nature Reports Climate Change*, **3** (0910), 115–16.

Molyneux, D. (2004), 'Neglected diseases but unrecognised successes – challenges and opportunities for infectious disease control', *Lancet*, **364**, 380–83.

Mooney, H.A., A. Duraiappah and A. Larigauderie (2013), 'Evolution of natural and social science interactions in global change research programs', *Proceedings of the National Academy of Sciences*, **110** (Early edition), 3665–72.

Moore, M. and F. Westley (2011), 'Surmountable chasms: Networks and social innovation for resilient systems', *Ecology and Society*, **16** (1), 5. Available at http://www.ecologyandsociety.org/vol16/iss1/art5/.

Moran M., J. Guzman, A-L. Ropars, A. McDonald, N. Jameson, B. Omune, S. Ryan and L. Wu (2009), 'Neglected disease research and development: How much are we really spending?', *PLoS Med*, **6** (2), e1000030.

Morozov, E. (2012), *The Net Delusion: The Dark Side of Internet Freedom*, New York: Public Affairs.

Morse, S.S. (2007), 'Global infectious disease surveillance and health intelligence', *Health Affairs*, **26** (4), 1069–77.

Moynihan, D.P. (2008), 'Learning under uncertainty: Networks in crisis management', *Public Administration Review*, March/April, 350–64.

Mykhalovskiy, E. and L. Weir (2006), 'The global public health intelligence network and early warning outbreak detection', *Canadian Journal of Public Health*, **97** (1), 42–4.

Nagurney, A., T. Wakolbinger and L. Zhao (2006), 'The evolution and emergence of integrated social and financial networks with electronic transactions: A dynamic supernetwork theory for the modeling, analysis, and computation of financial flows and relationship levels', *Computational Economics*, **27** (2–3), 353–93.

Nagy, B., J.D. Farmer, Q.M. Bui and J.E. Trancik (2013), 'Statistical basis for predicting technological progress', *PLoS One*, **8** (2), e52669.

Nasipour, S. (2012), 'US court scraps CFTC position limits rule'. Available at http://www.ft.com/cms/s/0/be191d8e-09a8-11e2-a424-00144feabdc0.html#axzz2RBEYLMuP (accessed 10 September 2013).

Nelson, M.C., K. Kintigh, D.R. Abbott and J.M. Anderies (2010), 'The cross-scale interplay between social and biophysical context and the vulnerability of irrigation-dependent societies: archaeology's long-term perspective', *Ecology and Society*, **15** (3), 31.

Nicholson, M. (2011), 'ICE to have authority to adjust softs trade prices' available at http://www.reuters.com/article/2011/06/27/us-ice-markets-softs-idUSTRE75Q5YE20110627 (accessed 10 September 2013).

Nilsson, M. and Å. Persson (2012), 'Can Earth system interactions be governed? Governance functions for linking climate change mitigation with land use, freshwater and biodiversity protection', *Ecological Economics*, **75**, 61–71.

Norberg, J. and G.S. Cumming (2008), *Complexity Theory for a Sustainable Future*, New York: Columbia University Press.

Norberg, J., M.C. Urban, M. Vellend, C.A. Klausmeier and N. Loeuille (2013), 'Eco-evolutionary responses of biodiversity to climate change', *Nature Climate Change*, **2**, 747–51.

Nuttall, M. (2012), 'Tipping points and the human world: Living with change and thinking about the future', *Ambio*, **41** (1), 96–105.

Nyström, M. and C. Folke (2001), 'Spatial resilience of coral reefs', *Ecosystems*, **4** (5), 406–17.

Nyström, M., A.V. Norström, T. Blenckner, M. de la Torre-Castro, J.S. Eklöf, C. Folke, H. Österblom, R.S. Steneck, M. Thyresson and M. Troell (2012), 'Confronting feedbacks in degraded marine ecosystems', *Ecosystems*, **15** (5), 695–710.

Oberthür, S. (2009), 'Interplay management: Enhancing environmental policy integration among international institutions', *International Environmental Agreements*, **9**, 371–91.

Oberthür, S. and O.S. Stokke (2011), *Managing Institutional Complexity – Regime Interplay and Global Environmental Change*, Cambridge, MA: The MIT Press.

OECD–FAO (2011), *Agricultural Outlook 2011–2020*, Organisation for Economic Co-operation and Development (OECD) Publishing and FAO (http://dx.doi.org/10.1787/agr_outlook-2011-en), 196.

Olsson, P. and V. Galaz (2012), 'Social–ecological innovation and transformation', in A. Nicholls and A. Murdoch (eds), *Social Innovation: Blurring Sector Boundaries and Challenging Institutional Arrangements*, Palgrave Macmillan: pp. 223–47.

Olsson, P., C. Folke and T.P. Hughes (2008), 'Navigating the transition to ecosystem based management of the Great Barrier Reef, Australia', *Proceedings of the National Academy of Sciences*, **105** (28), 9489–94.

Olsson, P., L.H. Gunderson, S.R. Carpenter, P. Ryan, L. Lebel and C. Folke (2006), 'Shooting the rapids: Navigating transitions to adaptive governance of social–ecological systems', *Ecology and Society*, **11** (1), 18.

Österblom, H. and U.R. Sumaila (2011), 'Toothfish crises, actor diversity and the emergence of compliance mechanisms in the Southern Ocean', *Global Environmental Change*, **21** (3), 972–82.

Österblom, H., S. Hansson, U. Larsson, O. Hjerne, F. Wulff et al. (2007), 'Human-induced trophic cascades and ecological regime shifts in the Baltic Sea', *Ecosystems*, **10** (6), 877–89.

Ostrom. E. (1990), *Governing the Commons – The Evolution of Institutions for Collective Action*, Cambridge, UK: Cambridge University Press.

Ostrom, E. (1998), 'A behavioral approach to the rational choice theory of collective action: Presidential Address, American Political Science Association', *American Political Science Review*, **92** (1), 1–22.

Ostrom, E. (1999), 'Coping with tragedies of the commons', *Annual Review of Political Science*, **2**, 493–535.

Ostrom, E. (2005), *Understanding Institutional Diversity*, Princeton: Princeton University Press.

Ostrom, E. (2007), 'A diagnostic approach for going beyond panaceas', *Proceedings of the National Academy of Sciences*, **104** (39), 15181–7.

Ostrom, E. (2010), 'Polycentric systems for coping with collective action and global environmental change', *Global Environmental Change*, **29**, 550–57.

Ostrom, E., J. Burger, C.B. Field, R.B. Noorgard and D. Policansky (1999), 'Revisiting the Commons: Local lessons, global challenges', *Science*, **284** (5412), 278–82.

Ostrom, V. (2000), 'Polycentricity, Part 2', in McGinnis, D (ed.), *Polycentric Games and Institutions: Readings from the workshop in political theory and policy analysis*, Ann Arbor, MI: University of Michigan Press, pp. 119–38.

Ostrom, V., C.M. Tiebout and R. Warren (1961), 'The organization of government in metropolitan areas: A theoretical inquiry', *American Political Science Review*, **55**, 831–42.

Pahl-Wostl, C., J. Sendzimir, P. Jeffrey, J. Aerts, G. Berkamp and K. Cross (2007), 'Managing change toward adaptive water management through social learning', *Ecology and Society*, **12** (2), 30.

Palkovacs, E.P., M.T. Kinnison, C. Correa, C.M. Dalton and A.P. Hendry (2012), 'Fates beyond traits: Ecological consequences of human-induced trait change', *Evolutionary Applications*, **5** (2), 183–91.

Park, H. (2003), 'Hyperlink network analysis: A new method for the study of social structure on the web', *Connections*, **25** (1), 49–61.

Patz, J., P. Daszak, G.M. Tabor, A. Aguirre, M. Pearl, J. Epstein, N.D. Wolfe, A.M. Kilpatrick, J. Foufopoulos, D. Molyneux and D.J. Bradley (2004), 'Unhealthy landscapes: Policy recommendations on land use change and infectious disease emergence', *Environmental Health Perspectives*, **1092** (10), 1092–8.

Perretti, C.T. and S.B. Munch (2012), 'Regime shift indicators fail under noise levels commonly observed in ecological systems', *Ecological Applications*, **22**, 1772–9.

Perrow, C. (1984), *Normal Accidents: Living with High-Risk Technologies*, Princeton: Princeton University Press.

Persson, L., M. Breitholtz, I. Cousins, C. de Wit, M. MacLeod and M. McLachlan (2013), 'Confronting unknown planetary boundary threats from chemical pollution', *Environmental Science and Technology*, pre-print preview: doi:10.1021/es402501c.

Pierre, J. and G.B. Peters (2005), *Governing Complex Societies: Trajectories and scenarios*, Hampshire: Palgrave Macmillan.

Pierson, P. (2000a), 'The limits of design: Explaining institutional origins and change', *Governance*, **13** (4), 475–99.

Pierson, P. (2000b), 'Path dependence, increasing returns, and the study of politics', *American Political Science Review*, **94** (2), 251–67.

Pierson, P. (2003), 'Big, slow-moving, and . . . invisible', in Mahoney James and Dietrich Rueschemeyer (eds), *Comparative Historical Analysis in the Social Sciences*, Cambridge: Cambridge University Press.

Pisano, Umberto, Andreas Endl and Gerald Berger (2012), 'The Rio+20 Conference 2012: Objectives, processes and outcomes', European Sustainable Development Network Quarterly Report No. 25. Vienna.

Plummer, R., B. Crona, D.R. Armitage, P. Olsson, M. Tengö and O. Yudina (2012), 'Adaptive comanagement: A systematic review and analysis', *Ecology and Society*, **17** (3), 11.

Poutanen, S.M., D.E. Low, B. Henry, S. Finkelstein, D. Rose, K. Green, R. Tellier, M.D.R. Draker, D. Adachi, M. Ayers, A.K. Chan, D.M. Skowronski, I. Salit, A.E. Simor, A.S. Slutsky, P.W. Doyle, M. Krajden, M. Petric, R.C. Brunham and A.J. McGeer (2003), 'Identification of severe acute respiratory syndrome in Canada', *The New England Journal of Medicine*, **348** (20), 1995–2005.

Price, M. (2011), 'Mifid II: in a nutshell'. Available at http://www.efinancialnews.com/story/2011-10-21/mifid-two-in-a-nutshell (accessed 10 September 2013).

Proctor, J.D. (2013), 'Saving nature in the Anthropocene', *Journal of Environmental Studies and Sciences*, **3** (1), 83–92.

Provan, K.G. and P. Kenis (2007), 'Modes of network governance: Structure, management and effectiveness', *Journal of Public Administration Research and Theory*, **18**, 229–52.

Rampton, R. (2011), 'ICE "loves" algo traders that trade 10 percent of softs'. Available at http://www.reuters.com/article/2011/03/16/us-futures-softs-idUSTRE72F99120110316 (accessed 10 September 2013).

Rau, G. (2011), 'CO_2 mitigation via capture and chemical conversion in seawater', *Environmental Science and Technology*, **45** (3), 1088–92.

Rau, G., E.L. McLeod and O. Hoegh-Guldberg (2012), 'The need for new ocean conservation strategies in a high-carbon dioxide world', *Nature Climate Change*, **2**, 720–24.

Raudsepp-Hearne, C., G.D. Peterson and E.M. Bennett (2010), 'Ecosystem service bundles for analyzing tradeoffs in diverse landscapes', *Proceedings of the National Academy of Sciences of the United States of America*, **107** (11), 5242–7.

Raworth, K. (2012), *A safe and just scape for humanity – can we live within the doughnut?*, Oxfam Discussion Papers, Oxfam.

Rayner, S., C. Heyward, T. Kruger, N. Pidgeon, C. Redgwell and J.Savulescu (2013), 'The Oxford principles', *Climatic Change* (January).

Redford, K.H., W. Adams, G. Mace, R. Carlson, S. Sanderson and S. Aldrich (2013), 'How will synthetic biology and conservation shape the future of nature?', Framing Paper, Wildlife Conservation Society/The Nature Conservancy (March 2013).

Reynolds, J. (2011), 'The regulation of climate engineering', *Law, Innovation and Technology*, **3** (1), 113–36.

Rigwell, A., J.S. Singarayer, A.M. Hetherington, and P.J. Valdes (2009), 'Tackling regional climate change by leaf albedo bio-geoengineering', *Current Biology*, **19** (2), 146–50.

Robbins, P. and S.A. Moore (2013), 'Ecological anxiety disorder: Diagnosing the politics of the Anthropocene', *Cultural Geographies*, **20** (1), 3–19.

Robock, A. (2008), '20 reasons why geoengineering may be a bad idea', *Bulletin of the Atomic Scientists*, **64** (2), 14–18.

Robock, A., M. Bunzl, B. Kravitz and G.L. Stenchikov (2010), 'Atmospheric science. A test for geoengineering?', *Science*, **327** (5965), 530–31.

Rockström, J. and L. Karlberg (2010), 'The quadruple squeeze: Defining the safe operating space for freshwater use to achieve a triply green revolution in the Anthropocene', *Ambio*, **39** (3), 257–65.

Rockström, J., W. Steffen, K. Noone, Å. Persson, F.S. Chapin, E. F. Lambin, T.M. Lenton, M. Scheffer, C. Folke, H.J. Schellnhuber, B. Nykvist, C.A. de Wit, T. Hughes, S. van der Leeuw, H. Rodhe, S. Sörlin, P.K. Snyder, R. Costanza, U. Svedin, M. Falkenmark, L. Karlberg, R.W. Corell, V.J. Fabry, J. Hansen, B. Walker, D. Liverman, K. Richardson, P. Crutzen and J.A. Foley (2009a), 'A safe operating space for humanity', *Nature*, **461**, 472–5.

Rockström, J., W. Steffen, K. Noone, Å. Persson, F.S. Chapin, E.F. Lambin, T.M. Lenton, M. Scheffer, C. Folke, H.J. Schellnhuber, B. Nykvist, C.A. de Wit, T. Hughes, S. van der Leeuw, H. Rodhe, S. Sörlin, P.K. Snyder, R. Costanza, U. Svedin, M. Falkenmark, L. Karlberg, R.W. Corell, V.J. Fabry, J. Hansen, B. Walker, D. Liverman, K. Richardson, P. Crutzen and J.A. Foley (2009b), 'Planetary boundaries: exploring the safe operating space for humanity', *Ecology and Society*, **14** (2), 32.

Roco, M.C. (2008), 'Possibilities for global governance of converging technologies', *Journal of Nanoparticle Research*, **10** (1), 11–29.

Royal Society (2009), *Geoengineering the Climate: Science, governance and uncertainty*, RS Policy document 10/09, London, UK.

Royal Society (2011), *Knowledge, Networks and Nations – Global Scientific Collaboration in the 21st century*, London, UK: The Royal Society.

Rubin, H. (2011), 'Future global shocks: Pandemics. Report from OECD/ IFP Project on "Future Global Shocks"', Rome: OECD.

Ruddiman, W.F. (2003), 'The anthropogenic greenhouse era began thousands of years ago', *Climatic Change*, **61** (3), 261–93.

Running, S.W. (2012), 'A measurable planetary boundary for the biosphere', *Science*, **337** (6101), 1458–9.

Sachs, J.D. (2012), 'From millennium development goals to sustainable development goals', *Lancet*, **379** (9832), 2206–11.

Sahgal, R. (2013), 'Forward Markets Commission lays down stringent norms for algorithm trades' available at http://articles. economictimes.indiatimes.com/2013-02-01/news/36684553_1_forward-markets-commission-algos-high-frequency-trades (accessed 10 September 2013).

Samper, C. (2009), 'Planetary boundaries: Rethinking biodiversity', *Nature Reports Climate Change*, **3** (0910), 118–19.

Sandström, A. and H. Ylinenpää (2012), 'Research, industry and public sector cooperation – a dynamic perspective', *International Journal of Innovation and Regional Development*, **4** (2), 144–59.

Scheffer, M. (2009), *Critical Transitions in Nature and Society*, Princeton: Princeton University Press.

Scheffer, M. and S. Carpenter (2003), 'Catastrophic regimes shifts in ecosystems: Linking theory to observation', *Trends in Ecology and Evolution*, **18** (12), 648–56.

Scheffer, M. and F. Westley (2007), 'The evolutionary basis of rigidity: Locks in cells, minds, and society', *Ecology and Society*, **12** (2), 36.

Scheffer, M., F. Westley and W. Brock (2003), 'Slow response of societies to new problems: causes and costs', *Ecosystems*, **6** (5), 493–502.

Scheffer, M., J. Bascompte, W.A. Brock, V. Brovkin, S.R. Carpenter, V. Dakos, H. Held, H.E. van Nes, M. Rietkerk and G. Sugihara (2009), 'Early-warning signals for critical transitions', *Nature*, **461** (7260), 53–9.

Scheffer, M., S.R. Carpenter, T.M. Lenton, J. Bascompte, W. Brock, V. Dakos, J. Van de Koppel, I.A. van de Leemput, S.A. Levin, E.H. van Nes, M. Pascual and J. Vandermeer (2012), 'Anticipating critical transitions', *Science*, **338** (6105), 344–8.

Schellnhuber, H-J, A. Block, M. Cassel-Gintz, J. Kropp, G. Lammel, W. Lass et al. (1997), 'Syndromes of global change', *GAIA*, **6**, 19–34.

Schlesinger, W.H. (2009), 'Planetary boundaries: Thresholds risk prolonged degradation', *Nature Reports Climate Change*, **3** (0910), 112–13.

Schlüter, M., R.R.J. McAllister, R. Arlinghaus and N. Bunnefeld (2012), 'New horizons for managing the environment: A review of coupled social–ecological systems modeling', *Natural Resource Modeling*, **25** (1), 219–72.

Schmidt, F. (2013), 'Governing planetary boundaries: Limiting or enabling conditions for transitions towards sustainability?', in L. Meuleman (ed.), *Transgovernance*, Berlin, Heidelberg: Springer Berlin Heidelberg, pp. 215–34.

Schoon, M. and C. Fabricius (2011), 'Synthesis: Vulnerability, traps, and transformations – long-term perspectives from archaeology', *Ecology and Society*, **16** (2), 24.

Scoones, I. (ed.) (2010), *Avian Influenza – Science, Policy and Politics*, London: Routledge.

Scott, K.N. (2013), 'International law in the Anthropocene: Responding to the geoengineering challenge', *Michigan Journal of International Law*, **34**, 309–58.

Seastedt, T.R., R.J. Hobbs and K.N. Suding (2008), 'Management of novel ecosystems: Are novel approaches required?' *Frontiers in Ecology and the Environment*, **6** (10), 547–53. doi:10.1890/070046.

Secretariat of the Convention on Biological Diversity (2012), *Cities and Biodiversity Outlook*. Montreal, Convention on Biological Diversity. 64pp.

Sethi, S.A. (2010), 'Risk management for fisheries', *Fish and Fisheries*, **11** (4), 341–65.

Seto, K.C., B. Güneralp and L.R. Hutyra (2012), 'Global forecasts of urban expansion to 2030 and direct impacts on biodiversity and carbon pools', *Proceedings of the National Academy of Sciences*, **109** (40), 16083–8.

Sheaff, R., L. Benson, L. Farbus, J. Schofield, R. Mannion and D. Reeves (2010), 'Network resilience in the face of health system reform', *Social Science and Medicine*, **70** (5), 779–86.

Six, K.D., S. Kloster, T. Ilyina, S.D. Archer, K. Zhang and E Maier-Reimer (2013), 'Global warming amplified by reduced sulphur fluxes as a result of ocean acidification', *Nature Climate Change*, advance online publication http://dx.doi.org/10.1038/nclimate1981.

Skelly D.K., L.N. Joseph, H.P. Possingham, L.K. Freidenburg, T.J. Farrugia, M.T. Kinnison and A.P. Hendry (2007), 'Evolutionary responses to climate change', *Conservation Biology*, **21** (5), 1353–5.

Slater, S. (2012), 'Barclays eyes quitting agricultural commods trading'. Available at http://www.reuters.com/article/2012/11/28/banks-barclays-commodities-idUSL5E8MSE7V20121128 (accessed 10 September 2013).

Snider, L. (2011), 'Criminalizing the algorithm? Stock Market crime in the 21st century', Plenary Address Delivered at the ANZ Conference of Criminology Geelong, Australia 28 September 2011.

Sornette, D. (2004), *Why Stock Markets Crash: Critical Events in Complex Financial Systems*, Princeton: Princeton University Press.

Specter, M. (2012), 'The climate fixers – is there a technological solution to global warming?', *New Yorker*. Available at http://www.new yorker.com/reporting/2012/05/14/120514fa_fact_specter?currentPage= all (accessed 14 September 2013).

Spiegel, J., S. Bennett, L. Hattersley, M.H. Hayden, P. Kittayapong, S. Nalim, D.N.C. Wang, E. Zielinski-Gutiérrez and D. Gubler (2005), 'Barriers and bridges to prevention and control of dengue: The need for a social–ecological approach', *EcoHealth*, **2** (4), 273–90.

Steffen, W., P.J. Crutzen and J.R. McNeill (2007), 'The Anthropocene: Are humans now overwhelming the great forces of nature?', *Ambio*, **36** (8), 614–21.

Steffen, W., J. Rockström and R. Constanza (2011), 'How defining planetary boundaries can transform our approach to growth', *Solutions*, **2** (3), 59–65.

Steffen, W., Å. Persson, L. Deutsch, J. Zalasiewicz, M. Williams, K. Richardson, C. Crumley, P. Crutzen, C. Folke, L. Gordon, M. Molina, V. Ramanathan, J. Rockström, M. Scheffer, H.J. Schellnhuber and U. Svedin (2011), 'The Anthropocene: from global change to planetary stewardship', *Ambio*, **40** (7), 739–61.

Steffen, W., A. Sanderson, P. Tyson, J. Jager, P. Matson, III, B. Moore, F. Oldfield, K. Richardson, H-J. Schellnhuber, B.L. Turner and R. Wasson (2004), *Global Change and the Earth System: A Planet under Pressure*, Heidelberg, Berlin: Springer Verlag.

Stern, N. (2007), *The Economics of Climate Change – The Stern Review*, Cambridge: Cambridge University Press.

Sterner, T., M. Troell, J. Vincent, S. Aniyar, S. Barrett, W. Brock et al. (2006), 'Quick fixes for the environment: Part of the solution or part of the problem?', *Environment: Science and Policy for Sustainable Development*, **48** (10), 20–27.

Stigebrandt, A. (2012), 'Evaluating geoengineering as a method to revive Baltic Sea dead zones', *Sea Technology Magazine*, December. Available at http://www.sea-technology.com/news/archives/2012/soapbox/soap box1212.php (accessed 10 September 2013).

Stodola, S. (2011), 'The Hathaway effect: Is Anne good for Berkshire?'. Available at http://www.thefiscaltimes.com/Blogs/Business-Buzz/2011/ 03/23/The-Hathaway-Effect-Is-Anne-Good-for-Berkshire.aspx#hZkHu tOAPJc88KCP.99 (accessed 10 September 2013).

Stone, C. (2004), 'Common but differentiated responsibilities in international law', *American Journal of International Law*, **98** (2), 276–301.

Struzik, E. (2013), 'Will bold steps be needed to save beleaguered polar bears?', Yale Environment 360. Available at http://e360.yale.edu/feature/will_bold_steps_be_needed_tosave_beleaguered_polar_bears/2618/(accessed 18 September 2013).

Sugiyama, M. and T. Sugiyama (2010), *Interpretation of CBD COP10 decision on geoengineering*. Socio-Economic Research Center Discussion Paper SERC10013. Central Research Institute of Electric Power Industry, Tokyo, Japan.

Sullivan, S. (2012), 'Banking nature? The spectacular financialisation of environmental conservation', *Antipode*, **45** (1), 198–217.

Tainter, J. (1988), *The Collapse of Complex Societies*, Cambridge, UK: Cambridge University Press.

Taleb, N. (2007), *The Black Swan: The Impact of the High Improbable*, New York: Random House.

Tavoni, A. (2013), 'Game theory: Building up cooperation', *Nature Climate Change*, **3** (9), 782–3. doi:10.1038/nclimate1962.

Teixeira, M.G., N. Da Conceicao, M. Costa, F. Barreto and M. Lima Barreto (2009), 'Dengue: Twenty-five years since reemergence in Brazil', *Cadernos de Saúde Pública*, **25**, 7–18.

Terazono, C. (2012), 'Commodity rally fails to lure back flows'. Available at http://www.ft.com/intl/cms/s/0/61d98b14-5d6c-11e1-869d-00144feabdc0.html#axzz2S8lyEiP9 (accessed 10 September 2013).

TheCityUK (2011), 'Commodities Trading March 2011', Financial Markets Series, London: Report by TheCityUK.

Tol, R. (2007), 'Europe's long-term climate target: A critical evaluation', *Energy Policy*, **35** (1), 424–32.

Tollefson, J. (2012), 'Ocean-fertilization project off Canada sparks furore', *Nature*, **490** (7421), 458–9.

Uhrqvist, O. and E. Lövbrand (2009), 'Seeing and knowing the Earth as a system – Tracing the history of the Earth System Science Partnership', Paper presented at the 2009 Amsterdam Conference on Human Dimensions of Global Environmental Change – Earth System Governance, People, places and the planet, 3 December 2009.

United Nations Conference on Trade and Development (UNCTD) (2009), *Development Impacts of Commodity Exchanges in Emerging Markets*, New York and Geneva. Report UNCTAD/DITC/COM/2008/9.

United Nations Conference on Trade and Development (UNCTD) (2011), *Price Formation in Financialized Commodity Markets – The Role of Information*, New York and Geneva. Report UNCTAD/GDS/2011/1.

United Nations Environment Programme (UNEP) (2012), GEO-5: Global

Environmental Outlook. Summary for Policy Makers. United Nations Environment Programme, Nairobi, Kenya.

United Nations High Level Task Force on the Global Food Security Crisis (2009), Progress Report April 2008–October 2009. United Nations, New York. Available at http://un-foodsecurity.org/sites/default/files/09progressreport.pdf (accessed 10 September 2010).

United States Commodity Futures Trading Commission and US Securities and Exchange Commission (2010), Preliminary Findings Regarding the Market Events of 6 May 2010 – Report of the Staffs of the CFTC and SEC to the Joint Advisory Committee on Emerging Regulatory Issues (May).

United States Government Accountability Office (2011), *Climate Engineering: Technical status, future directions, and potential responses*, Washington, DC: United States Government Accountability Office, p. 135.

van Baalen, J.P. and P.C. van Fenema (2009), 'Instantiating global crisis networks: The case of SARS', *Decision Support Systems*, **47** (4), 277–86.

Vandegrift, K. and S. Sokolow (2010), 'Ecology of avian influenza viruses in a changing world', *Annals of the New York Academy of Sciences*, 113–28.

VanderZwaag, D. (1999), 'The precautionary principle in environmental law and policy: Elusive rhetoric and first embraces', *Journal of Environmental Law and Practice*, **8** (3), 355–75.

Vasconcelos, V.V., F.C. Santos and J.M. Pacheco (2013), 'A bottom-up institutional approach to cooperative governance of risky commons', *Nature Climate Change*, **3** (9), 797–801. doi:10.1038/nclimate1927.

Victor, D.G. (2010), *Global Warming Gridlock: Creating more effective strategies for protecting the planet*, Cambridge, UK: Cambridge University Press.

Vidal, J. (2012), 'Bill Gates backs climate scientists lobbying for large-scale geoengineering'. Available at http://www.guardian.co.uk/environment/2012/feb/06/bill-gates-climate-scientists-geoengineering (accessed 17 September 2013).

Vidas, D. (2011), 'The Anthropocene and the international law of the sea', *Philosophical Transactions of the Royal Society of London*, Series A, Mathematical, Physical, and Engineering Sciences, **369** (1938), 909–25. doi:10.1098/rsta.2010.0326.

Virgoe, J. (2008), 'International governance of a possible geoengineering intervention to combat climate change', *Climatic Change*, **95** (1–2), 103–19.

Walker, B., S. Barrett, S. Polasky, V. Galaz, C. Folke, G. Engström, F. Ackerman, K. Arrow, S. Carpenter, K. Chopra, G. Daily, P. Ehrlich,

T. Hughes, N. Kautsky, S. Levin, K-G. Mäler, J. Shogren, J. Vincent, T. Xepapadeas and A. de Zeeuw (2009), 'Looming global-scale failures and missing institutions', *Science*, **325** (5946), 1345–6.

Walker, B.H., N. Abel, J.M. Anderies and P. Ryan (2009), 'Resilience, adaptability, and transformability in the Goulburn-Broken Catchment, Australia', *Ecology and Society*, **14** (1), 12.

Wallace, D.W.R., C.S. Law, P.W. Boyd, Y. Collos, P. Croot, K. Denman, P.J. Lam, U. Riebesell, S. Takeda and P. Williamson (2010), *Ocean Fertilization. A Scientific Summary for Policy Makers*. IOC/UNESCO, Paris (IOC/BRO/2010/2).

Wallinga, J. and Teunis, P. (2004), 'Different epidemic curves for severe acute respiratory syndrome reveal similar impacts of control measures', *American Journal of Epidemiology*, **160** (6), 509–16.

Wall Street Journal, The (2012), 'The global doomsayers' ever-changing story', *The Wall Street Journal*, 15 June 2012. Available at http://online.wsj.com/news/articles/SB10001424052702303901504577460900066999454 (accessed 2 December 2013).

Wang, R., J.A. Dearing, P.G. Langdon, E. Zhang, X. Yang, V. Dakos and M. Scheffer (2012), 'Flickering gives early warning signals of a critical transition to a eutrophic lake state', *Nature*, **492**, 419–22.

Westley, F. (2013), 'Social innovation and resilience: How one enhances the other', Stanford Social Innovation Review, Summer 2013 (Supplement 'Innovation for a Complex World'), 6–8.

Westley, F., P. Olsson, C. Folke, T. Homer-Dixon, H. Vredenburg, D. Loorbach, J. Thompson, M. Nilsson, E. Lambin, J. Sendzimir, B. Banerjee, V. Galaz and S. van der Leeuw (2011), 'Tipping toward sustainability: emerging pathways of transformation', *Ambio*, **40**, 762–80.

Wiggins, A. and K. Crowston (2011), 'From conservation to crowdsourcing: A typology of citizen science', Conference Paper presented at the 44th Hawaii International Conference on System Science (HICSS-44).

Wild, M., A. Ohmura and K. Makowski (2007), 'Impact of global dimming and brightening on global warming', *Geophysical Research Letters* 34:L04702. Available at http://dx.doi.org/doi:10.1029/2006GL028031.

Williamson, P., R.T. Watson, G. Mace, P. Artaxo, R. Bodle, V. Galaz, A. Parker, D. Santillo, C. Vivian, D. Cooper, J. Webbe, A. Cung and E. Woods (2012), *Impacts of Climate-Related Geoengineering on Biological Diversity. Part I of: Geoengineering in Relation to the Convention on Biological Diversity: Technical and Regulatory Matters*, Montreal: Secretariat of the Convention on Biological Diversity, Technical Series No. 66, p. 152.

World Economic Forum (2013), *Global Risks 2013 – Eighth Edition*, Geneva: World Economic Forum.

World Federation of Exchanges (2011), 'Improving structure, promoting quality', Annual Report 2011, Paris: World Federation of Exchanges.

World Health Organization (WHO) (2005), *WHO Global Influenza Preparedness Plan – The role of WHO and recommendations for national measures before and during pandemics*, Geneva: World Health Organization.

Yamagishi, T. and K.S. Cook (1993), 'Generalized exchange and social dilemmas', *Social Psychology Quarterly*, **56**, 235–48.

Yamey, G. (2002), 'The world's most neglected diseases', *British Medical Journal*, **325**, 176–7.

Young, O.R. (2008), 'Building regimes for socioecological systems', in O.R. Young, L.A. King and H. Schroeder (eds), *The Institutional Dimensions of Global Environmental Change: Principal findings and future directions*, Cambridge, MA: The MIT Press, pp. 115–44.

Young, O.R. (2010), *Institutional Dynamics – Emergent Patterns in International Environmental Governance*, Cambridge, MA: The MIT Press.

Young, O.R. (2010), 'Institutional dynamics: Resilience, vulnerability and adaptation in environmental and resource regimes', *Global Environmental Change*, **20** (3), 378–85.

Young, O.R. (2012), 'Arctic tipping points: governance in turbulent times', *Ambio*, **41** (1), 75–84.

Young, O.R., L.A. King and H. Schroeder (eds) (1998), *The Institutional Dimensions of Global Environmental Change: Principal findings and future directions*, Cambridge, MA: The MIT Press.

Young, O., F. Berkhout, G.C. Gallopin, M.A. Janssen, E. Ostrom and S. van der Leeuw (2006), 'The globalization of socio-ecological systems: An agenda for scientific research', *Global Environmental Change*, **16** (3), 304–16.

Zalasiewicz, J., M. Williams, W. Steffen and P. Crutzen (2010), 'The new world of the Anthropocene', *Environmental Science and Technology*, **44** (7), 2228–31.

Zalasiewicz, J., M. Williams, A. Smith, T.L. Barry, A.L. Coe, P.R. Bown, P. Brenchley, D. Cantrill, A. Gale, P. Gibbard, F.J. Gregory, M.W. Hounslow, A.C. Kerr, P. Pearson, R. Knox, J. Powell, C. Waters, J. Marshall, M. Oates, P. Rawson and P. Stone (2008), 'Are we now living in the Anthropocene?', *GSA Today*, **18** (2), 4.

Index